THE GOVERNANCE OF LARGE TECHNICAL SYSTEMS

Large technical systems (LTSs) such as transportation, telecommunications, energy supply and urban utility networks present particular problems of governance. Their complexity calls for constant innovation in firm management and industrial organization, while their key role in modern societies and their frequent monopolistic nature necessitate close public surveillance.

This work examines the governance of LTSs at firm, industry and government levels, and the interactions between the systems and society. In particular, international contributors explore the implications of major technological, economic and social changes during the last twenty years for traditional forms of LTS governance. Their research is centred around the following themes:

- traditional forms of governance
- new regulatory challenges
- the governability of complex technologies
- conceptual issues related to the governance of interorganizational networks.

The Governance of Large Technical Systems combines international case studies with conceptual research, offering a wide-ranging discussion of governance issues for the large technical systems so essential to industrialized and industrializing countries.

Olivier Coutard is a researcher with France's Centre National de la Recherche Scientifique (CNRS) and co-ordinator of their Research Group on Networks. He is a member of the editorial boards of *Flux, Cahiers scientifiques internationaux Réseaux et Territoires* and of the *Journal of Urban Technology*.

ROUTLEDGE STUDIES IN BUSINESS ORGANIZATIONS AND NETWORKS

1 DEMOCRACY AND EFFICIENCY IN THE ECONOMIC ENTERPRISE
Edited by Ugo Pagano and Robert Rowthorn

2 TOWARDS A COMPETENCE THEORY OF THE FIRM
Edited by Nicolai J. Foss and Christian Knudsen

3 UNCERTAINTY AND ECONOMIC EVOLUTION
Essays in honour of Armen A. Alchian
Edited by John R. Lott Jr

4 THE END OF THE PROFESSIONS?
The restructuring of professional work
Edited by Jane Broadbent, Michael Dietrich and Jennifer Roberts

5 SHOPFLOOR MATTERS
Labor-management relations in twentieth-century American manufacturing
David Fairris

6 THE ORGANISATION OF THE FIRM
International business perspectives
Edited by Ram Mudambi and Martin Ricketts

7 ORGANIZING INDUSTRIAL ACTIVITIES ACROSS FIRM BOUNDARIES
Anna Dubois

8 ECONOMIC ORGANISATION, CAPABILITIES AND COORDINATION
Edited by Nicolai Foss and Brian J. Loasby

9 THE CHANGING BOUNDARIES OF THE FIRM
Explaining evolving inter-firm relations
Edited by Massimo G. Colombo

10 AUTHORITY AND CONTROL IN MODERN INDUSTRY
Theoretical and empirical perspectives
Edited by Paul L. Robertson

11 INTERFIRM NETWORKS
Organization and industrial competitiveness
Edited by Anna Grandori

12 PRIVATIZATION AND SUPPLY CHAIN MANAGEMENT
Andrew Cox, Lisa Harris and David Parker

13 THE GOVERNANCE OF LARGE TECHNICAL SYSTEMS
Edited by Olivier Coutard

THE GOVERNANCE OF LARGE TECHNICAL SYSTEMS

Edited by Olivier Coutard

ROUTLEDGE

London and New York

First published 1999
by Routledge
11 New Fetter Lane, London EC4P 4EE

Simultaneously published in the USA and Canada
by Routledge
29 West 35th Street, New York, NY 10001

Typeset in Garamond by Keystroke, Jacaranda Lodge, Wolverhampton
Printed and bound in Great Britain by Biddles Ltd, Guildford and King's Lynn

British Library Cataloguing in Publication Data
A catalogue record for this book is available from the British Library

Library of Congress Cataloging in Publication Data
The governance of large technical systems / edited by Olivier Coutard.
Includes bibliographical references and index.
1. Public utilities – Management. 2. Transportation – Management.
3. Telecommunication – Management. 4. Corporate governance.
I. Coutard, Olivier.
HD2763.G64 1999
363.6 – dc21 98–33941

ISBN 0-415-19603-5

CONTENTS

ILLUSTRATIONS

Figures

Tables

CONTRIBUTORS

Janet Abbate is lecturer in the Department of History at the University of Maryland. Her work on LTSs has focused on computer networks and she recently finished a book, *Inventing the Internet*, forthcoming at MIT Press in 1999.

Konstantinos Chatzis is a researcher at Laboratoire Techniques Territoires Sociétés. He is interested in the history of science and technology and more specifically in the evolution of large socio-technical systems since the middle of the nineteenth century. He is the author of over thirty articles and of a book on the history of sewerage systems. Publications on the governance of large technical systems and engineers include: *Les dessous de la ville. L'ingénieur, l'égout et la rame*, Paris: L'Harmattan (forthcoming); and 'Looking for standards: French engineers and work measurement in the 1950s', *History and Technology* (forthcoming).

Olivier Coutard holds a full-time research position with CNRS, where he researches on the social and urban aspects of large technical systems. He is also coordinator of the CNRS Research Group on Networks. His publications on the governance of large technical systems include: 'Economics of Grid Systems in Reconfiguration: Deregulation in the Electricity Supply Industry in Europe and the United States', in Jane Summerton (ed.) *Changing Large Technical Systems*, Boulder (CO): Westview (1994); and 'Quinze ans de déréglementation des services publics: les exemples de l'électricité et des télécommunications', *Sociologie du travail* 39, 1 (1997). He can be contacted by e-mail (coutard@enpc.fr).

Stephen Graham is a reader in the Centre for Urban Technology (CUT), University of Newcastle. His research centres on cities and infrastructure, urban theory, and telecommunications and city development and planning. His recent books include: *Telecommunications and the City: Electronic Spaces, Urban Places* (with S. Marvin), London: Routledge (1996); and *Managing Cities: The New Urban Context* (co-editor, with P. Healey, S. Cameron, S. Davoudi and A. Madani-Pour), London: Wiley (1995).

Alain Gras is professor of sociology and anthropology at the Department of Philosophy, University of Paris 1, where he heads the Centre d'Etude des Techniques, des Connaissances et des Pratiques (CETCOPRA). He is interested in theoretical issues in technology studies (the history of technology as a non-linear process, a critical approach to the theory of evolution), based on applied research in aeronautics, transportation systems, automation and simulators. His publications include: *Sociologie des Ruptures. Les pièges du temps en sciences sociales*, Paris: PUF (1979); *Sociologie des Techniques de la vie quotidienne* (ed., with V. Scardigli and B. Joerges), Paris: L'Harmattan (1992); *Grandeur et dépendance. Sociologie des macro-systèmes techniques*, Paris: PUF (1993); and *Les macro-systèmes techniques*, Paris: PUF (1997).

Pascal Griset is professor of history at the University of Paris 4, where he heads the Research Centre on the History of Innovation. His research deals with the economic and technological history of information systems, and with the history of innovations in France and in the US. Two of his books were awarded French academic prizes: *Les révolutions de la communication (XIXème–XXème siècles)*, Paris: Hachette Supérieur (1992); and *Technologie, entreprise et souveraineté: les télécommunications transatlantiques de la France*, Paris: Institut d'Histoire de l'Industrie and Editions Rive Droite (1996).

Reiner Grundmann is a lecturer at Aston Business School, Aston University, UK. He has just completed a book on the CFC–zone controversy in an international perspective entitled *Transnationale Umweltpolitik zum Schutz der Ozonschicht*, Frankfurt: Campus (forthcoming). Previous publications include *Marxism and Ecology*, Oxford: Oxford University Press (1991). His research interests include political sociology, ecology, sociology of science and technology, and risk and uncertainty.

Simon Guy is a lecturer in the Built Environment in the Department of Architecture at the University of Newcastle, UK. His research interests revolve around the social production of the material environment. He has undertaken research into a wide spectrum of urban technology issues funded by the Economic and Social Research Council, the Engineering and Physical Research Council and the European Union. Recent publications include: 'Privatised utilities and regional governance: the new regional managers?' (with S. Graham and S. Marvin), *Regional Studies* 30, 8 (1996); and 'Transforming urban infrastructure provision: the emerging logic of demand side management' (with S. Marvin), *Policy Studies* 17, 2 (1996).

Bernward Joerges is a researcher at the Wissenschaftszentrum Berlin für Sozialforschung (WZB) and heads the research group 'Metropolitan studies'. He is also professor of sociology at the Technische Universität Berlin. Recent publications include: *Technik ohne Grenzen* (editor, with Ingo Braun), Frankfurt am Main: Suhrkamp (1994); and *Technik: Körper der Gesellschaft*, Frankfurt am Main: Suhrkamp (1996).

Arne Kaijser is professor at the Department of History of Science and Technology, Royal Institute of Technology, Stockholm. His main research interest is comparative studies of the historical development of large technical systems. His most recent books are: *Nordic Energy Systems. Historical Perspectives and Current Issues* (co-edited with Marika Hedin), Canton (MA): Science History Publications (1995); and *Den konstruerade världen. Tekniska system i historiskt perspektiv* (co-edited with Pär Blomkvist), Stockholm: Brutus Östlings Bokförlag Symposion (1998). He was research fellow at the Technical University of Delft in 1993/94.

Francis McGowan is a lecturer in politics in the School of European Studies and a research fellow in the Sussex European Institute, both Sussex University. His research covers the public utilities and international regulatory issues. Recent publications include: *European Energy Policy in a Changing Environment* (ed.), Berlin: Physica Verlag (1996); and 'La réforme des services publics au Royaume-Uni: privatisation, réglementation et liberalisation' (with A. Holmes), *Sciences de la Société* 43 (1997).

Donald MacKenzie holds a personal chair in Sociology at Edinburgh University, where he has taught since 1975. He is the author of *Statistics in Britain, 1865–1930: The Social Construction of Scientific Knowledge*, Edinburgh: Edinburgh University Press (1981); *Inventing Accuracy: A Historical Sociology of Nuclear Missile Guidance*, Cambridge (MA): MIT Press (1990); and of *Knowing Machines: Essays on Technical Change*, Cambridge (MA): MIT Press (1996). The second of these books won the Ludwik Fleck prize of the Society for Social Studies of Science, and was joint winner of the 1993 Robert K. Merton Award of the American Sociological Association.

Simon Marvin is director of the Centre for Urban Technology at the University of Newcastle. He has research interests in the relationships between infrastructure networks and urban and regional governance in developed and developing cities. His research programme has received funding from the UK Research Councils, the European Commission, private companies and governmental organizations. He has published widely on the changing relations between cities and technical networks, and was co-author with Stephen Graham of *Telecommunications and the City*, London: Routledge (1997).

Jean-Marc Offner is a researcher at the Laboratoire Techniques Territoires Sociétés (LATTS), and professor at the Ecole Nationale des Ponts et Chaussées, where he heads the planning studies department. He is also director of the Groupement de recherche 'Réseaux' (CNRS). He has recently co-edited with Denise Pumain *Réseaux et Territoires: significations croisées*, La Tour d'Aigues: Editions de l'Aube (1996).

Jane Summerton is a researcher at the Department of Technology and Social Change, University of Linköping, Sweden. As a sociologist she analyses actor dynamics in sociotechnical systems. Among many publications, she recently edited the international anthology *Changing Large Technical Systems*, Boulder (CO): Westview (1994).

Joel A. Tarr is the Richard S. Caliguiri Professor of Urban and Environmental History and Policy at Carnegie Mellon University. His main research interests are in the history of urban technological systems and of urban environmental pollution. He has published on transportation, energy and communications systems and on issues of air, land and water pollution. His latest book is *The Search for the Ultimate Sink: Urban Pollution in Historical Perspective*, Akron (OH): University of Akron Press (1996).

ABOUT THE BOOK

This book is about large technical systems (LTSs), the capital-intensive and often complex socio-technical organizations that have been deployed to provide for the crucial needs of modern industrial societies. The archetypes of LTSs are the transportation, telecommunications, energy supply and urban utility networks that have been developed from the second half of the nineteenth century. Because they are complex, these systems have traditionally fostered innovations in industrial organization and management. Because they are essential to the functioning of modern territories and economies, and because they are traditionally organized as monopolies they were always placed under close public surveillance. More specifically, the book is about the *governance* of LTSs – at the level of individual firms (management), of economic sectors (industrial organization and public regulation) and of the broader interactions between these systems and 'society at large'.

LTSs have been affected by major technological, economic and social changes during the past two decades (including: innovations in information technologies; the calling into question of hierarchical structures, both in industry and in government; the growing distrust of states' intervention in the economy; the decline of popular enthusiasm for these technologies; and the diverging logics and interests of industry and of nation-states). The book explores the implications of these changes on traditional forms of LTS governance.

It contains an introduction, and thirteen chapters written by internationally-acknowledged social science scholars. The chapters are organized in four parts respectively addressing traditional forms of governance; new regulatory challenges; the governability of complex technologies; and conceptual issues related to the governance of inter-organizational networks. By its combination of case studies and conceptual developments, the book offers a wide-ranging discussion of governance issues in large technical systems – issues which are of primary importance in industrialized and industrializing countries.

ACKNOWLEDGEMENTS

The chapters of this book have been selected among a broader collection of papers presented and discussed at a roundtable conference on the governance of large technical systems, which was held in Autun (France) in September 1995. The purpose of the conference was to gather two groups of scholars: the French Research Group on Networks (Groupement de recherche 'Réseaux', CNRS) – which develop a territorial approach in the study of 'urban technical networks' – and the group of scholars involved in the international conference series on 'the development of large technical systems' – who emphasize the interactions between social systems and technological systems.

After the conference the papers were reviewed, sometimes substantially, and updated for the book. Bernard Barraqué (CNRS), Ingo Braun (then at the Wissenschaftszentrum Berlin), Patrice Flichy (CNRS), Stephen Graham (University of Newcastle), Alain Gras (University of Paris 1), Dominique Lorrain (CNRS), Denise Pumain (University of Paris 1), Gene Rochlin (University of Berkeley), Stephen Salsbury (University of Sydney), and Jane Summerton (University of Linköping) have contributed to the selection and review of the papers that form the chapters of this book. Bernward Joerges (Wissenschaftszentrum Berlin), Renate Mayntz (Max Planck Insitut für Gesellschaftsforschung Köln) and Thomas P. Hughes (University of Pennsylvania) have brought precious advice and support to the project. Annie Gaumain, Oleg Kraïowsky-Koukel, and Neil O'Brien have greatly contributed to the making of this book.

Many public and private institutions in France have provided intellectual, material and financial help for this project: the Centre National de la Recherche Scientifique (Département des Sciences de l'Homme et de la Société); the Ministère de l'Equipement, des Transports et du Logement (Direction de la Recherche et des Affaires Scientifiques et Techniques); the Ecole Nationale des Ponts et Chaussées; the Délégation à l'Aménagement du Territoire et à l'Action Régionale; the Programme Interdisciplinaire de Recherche sur les Villes (CNRS), and its director Gabriel Dupuy; the Commissariat Général du Plan; France Télécom (Service des Etudes

Economiques); La Poste (Direction de la Stratégie); Cofiroute; Gaz de France (Délégation à la Stratégie); Electricité de France (Mission Prospective).

Finally, I would like to express my gratitude to Jane Summerton (again) and Jean-Marc Offner, the director of the Research Group on Networks, for their constant and friendly support.

Olivier Coutard
Marne-la-Vallée, July 1998

1

INTRODUCTION

The evolving forms of governance of large technical systems

Olivier Coutard

In memory of Stephen M. Salsbury

The objects studied in this book are the capital intensive and often complex socio-technical organizations that have been deployed to meet the crucial needs of modern industrial societies. These organizations include telecommunications, transportation, and energy supply systems, as well as urban technical networks such as water supply or sewerage systems. Because these systems include complex technological artefacts, they are often viewed as 'engineers' stuff', not worth the interest of the social sciences. One salient and long-lasting exception to this embargo concerns economic studies, because these systems necessitate massive investments, and for this reason are traditionally organized as territorial monopolies. But it has long been considered that the social implications of these systems were limited to the economic regulation of the resulting monopolies. This book, together with a number of others (see in particular Hughes 1983, 1989, Mayntz and Hughes 1988, La Porte 1991, Summerton 1994a, and Braun and Joerges 1994b), takes a different stance, by assuming that the relations between these socio-technical systems and 'society at large' are essential to understand the evolution – the coevolution, one might say – of both the systems *and* the society.

The subject of the book is to examine how these systems have been governed in the past and how they may be governed in the future, at the level of the individual firms (management); of economic sectors (industrial organization and public regulation); and of the broader interactions of these systems with society at large. Previous books sharing the same perspective – the so-called 'systems perspective' – have obviously been concerned with control, management and regulatory issues; this book adds to previous findings and systematically develops some of the major aspects of the problem.

The book is thus about the 'governance' of large technical systems (LTSs). The use of the currently fashionable notion of governance calls for some justification. Indeed governance is a polysemous concept; as Jouve, Lefèvre and Offner point out:

> Firstly governance . . . may designate either an action or an organization, either a process or an institution. Secondly . . . governance simultaneously refers to general categories (all forms of management of transactions; local government) and to particular 'species' (network transactions, between hierarchies and markets; fragmented local institutions). Finally the fact that the notion originates from two distinct disciplinary fields – economics and management on the one hand, political science on the other – adds up to the confusion . . .
>
> (Jouve, Lefèvre and Offner 1995: 10)

I shall (boldly) argue below that I have retained the word 'governance' precisely for its polysemy! For, indeed, the subject of this book mirrors the wide scope of the concept of governance: it is concerned both with the economics and with the politics of LTSs; it deals with the general issue of the control of LTSs as well as with its more specific decentralized forms; and it also examines tensions between actions and institutions. I should add that it also deals with the interactions between the (various forms and levels of) governance of technological systems and societal governance patterns.

This book thus makes pragmatic use of the notion of governance; it is not primarily concerned with the clarification of that notion. However it would be a mistake to assume that the relevance of the notion of governance for the purposes of this book derives from its poor semantic specification. On the contrary, the scope of the notion emphasizes one important assumption of this book: namely, that the three levels of governance (the level of LTS firms' management, the level of LTS sectors' industrial organization and public regulation, and the level of LTS–society interactions) are closely linked. In this respect only can the chapters of this book provide some insights into the general problem of societal governance. Besides, as Bernward Joerges suggests elsewhere, 'vague' terms can act as useful gateways between intellectual cultures, disciplines or approaches.[1]

That a governance perspective is heuristic for our purpose is also suggested by the numerous affinities it has with the notion of network, which is one of the central themes discussed in the book (in particular in the three contributions that form the last part of this volume). On the one hand, governance in its more general meaning (suggesting ideas of control or steering) is questioned by the emergence of networks (and the idea of self-organization suggested by the notion of network). Thus defined, the problem (the challenge) of the governance of networked organizations is a central concern of this book. On the other hand, the affinity with the more specified notions of governance (suggesting ideas of fragmented control) is obvious – whether one refers to 'policy networks' (Marin and Mayntz 1991), to networks as a specific form of (the management of) economic or social transactions (Thompson *et al.* 1991) or to networks as sets of interorganizational relations (see the discussion below). Both the notions of governance and of network emphasize the rise of multi-actor system builders (see Hughes 1996); of decentralized LTSs (see Braun and Joerges 1994a); of multi-organizations, multi-stakes sytems – which this book is about.

Let us now consider the organization of the book, and the thirteen contributed chapters it includes.

New insights on the development of large technical systems

The focus, in the systems approach, on coevolutionary mechanisms has proved very useful in analysing the expansionist dynamics of LTSs, from local to regional to large-scale, integrated, hierarchical systems (Mayntz and Hughes 1988). On the one hand, the importance of 'load factor'[2] and 'economic mix'[3] has been repeatedly demonstrated: these economic factors act as major forces in the spatial growth of power systems, and of LTSs in general. On the other hand, the problems raised by the expansion of systems across institutional (political, jurisdictional) borders have also been borne out.[4]

None the less, the development of LTSs does not necessarily imply spatial expansion. Offner (in this volume) distinguishes between 'extensive' and 'intensive' development patterns. Although extensive development (the interconnection of heterogeneous networks to form a new, 'hybrid' system) usually involves the upsizing of the resulting system, it sometimes occurs that the system is actually downsized (as the Paris region express subway system described by Offner exemplifies). 'Intensive' development consists of the 'densification' of the network, through the integration of flows of different nature;[5] it does not necessarily involve any change in the system's size.

Moreover, the development of LTSs cannot be reduced to their spatial dimensions. Sometimes the 'largeness' of an LTS does not even refer to its size at all. Such is the case for the sewerage systems considered by Chatzis (this volume): if sewerage systems readily qualify as 'large' technical systems, this is not because of the large size of individual systems, but because these systems are designed and operated according to standard rules and practices endowed with large-scale validity. Rather than by technical devices, Chatzis suggests that sewerage systems are interconnected by immaterial devices such as standards and professional associations.

That the development of LTSs is not always a spatial phenomenon is also clear from a recent collection of essays on reconfigurations in LTSs (Summerton 1994a). In her introduction to that book, Jane Summerton identifies three types of reconfiguration: the territorial expansion and interconnection of similar systems across political borders; the crossing of functional system boundaries; and the reorganization of former monopoly systems into new configurations based on principles of competition and free access (Summerton 1994b: 5–6). Only the first type of reconfiguration is primarily spatial.

The chapters by Chatzis, Griset, Kaijser and Tarr in Part 1 of this volume all concern the development of LTSs; yet they do not focus on spatial-economic dynamics. Rather they explore other dynamics, and thus offer new insights into the traditional forms of governance of LTSs.

Joel Tarr describes the evolution of the manufactured gas system in the US, and its transformation into a natural gas system. His study places more emphasis on technological and regulatory aspects, and less on organizational and managerial ones. Indeed it is striking that the transition to natural gas, a major reconfiguration,[6] was achieved without any significant change in the industry structure: technological change was implemented from within existing companies. Parallel to his discussion of technological choices throughout the US gas history, Tarr analyses the importance of municipal, state and ultimately federal regulation in shaping the natural gas system. Even though he emphasizes the 'corporate structure' of the US gas industry, Tarr shows how all significant steps in the transition were supervised by regulatory authorities. This is not to suggest, however, the existence of one wise regulator preventing the industry from going astray; on the contrary, Tarr provides striking examples of regulators faced with similar problems in similar contexts, and reaching very different conclusions.

Arne Kaijser's study also deals with the transition from manufactured to natural gas, but in a very different context, and from a different perspective. Kaijser examines the consequences of the discovery of a huge gas field in the Netherlands in the late 1950s. He focuses primarily on the new gas 'regime' ('the legal framework, organizational structure and rules of the game' within the industry) that was designed to derive the highest benefits from the discovery. He analyses the 'game' among political and corporate actors. But his study goes beyond the influence of individual actors: by describing the leading role played by the successive Dutch ministers for economic affairs, Kaijser clearly shows that public leadership in this matter was ongoing, even if consensus was frequently obtained due to the fact that political and corporate actors shared a number of common values (including the value of consensus).

In a chapter on telegraphy, Pascal Griset examines the competition between wireless and cable systems in the early twentieth century. Griset's account of the expansion of wireless telegraphy refers repeatedly to space. But the notion of space at stake is not 'economic space' that can be interpreted in terms of load factor and economic mix; it is 'geopolitical' or strategic space – a peacetime battleground for nations in search of international influence. Griset also sheds new light on the attitude of politicians towards technology, in that particular case, in favour of centralized, big technologies.

In dealing with systems of a local nature, sewerage systems, Konstantinos Chatzis is at variance with most LTS research.[7] In his careful study of engineers' *design and operation practices*, Chatzis emphasizes yet another dimension of system control. Analysing the role of (communities of) engineers in the design of supposedly technical rules, Chatzis describes how these rules in fact incorporate social norms and values. Interestingly enough, although there is no explicit mention of public oversight in Chatzis's study, the engineers responsible for the conception of design and operation rules in France are all civil servants in the national roads administration ('corps des ponts et chaussées'), and therefore all strongly committed to the 'public interest'; and the implementation of these rules was definitely a top-down process, with the rules being elaborated at national level and then forced upon local authorities and technical bodies through legislation.

These four case studies remind us that the development of these systems does not boil down to spatial expansion fuelled by technical-economic considerations and (possibly) impeded by political parochialism (see also e.g. Salsbury 1988, Coutard 1995, Offner and Pumain 1996). This, of course, has always been common knowledge within the systems approach to technology. But the dynamics exposed here (the re-use of an existing system, the role of moral values, of 'geopolitics' or of 'technical-scientific referentials') have seldom been explicitly dealt with; they have more often been mentioned as part of the 'cultural' context in which LTSs develop.

A changing background

These historical case studies, despite their diversity,[8] share a common 'background' that can be characterized by four important features: (1) the predominance of hierarchical organizations, both in government and in industry; (2) a strong influence of national public (or regulatory) authorities; (3) general acceptance of LTSs by the public;[9] and (4) shared interests between LTS companies and national governments.

This background was substantially challenged over the past two decades. First, the economic performance of hierarchical systems of production has been increasingly questioned. The most frequent critics have emphasized the lack of responsiveness of hierarchical systems to consumer needs, and their poor achievements in terms of technological and organizational innovation. In contrast, decentralized systems are perceived as flexible, innovative and efficient. More generally, mass production is losing ground to just-in-time, client-oriented and rapidly-evolving manufacturing processes, calling for more and more complex forms of coordination within and among firms, and within and among markets (Veltz 1996). Hierarchies in government were also criticized. In particular, the role of national governments as the central public actors and as the ultimate judges of the 'general interest' was challenged.

Second, these critics of 'Fordian' forms of industrial organization were accompanied by a growing distrust of governmental action, and in particular of public intervention in economic activities. LTSs were therefore criticized both as hierarchical organizations and as strongly regulated economic activities. In many instances they were perceived as symbols of the shortfalls of economic policy. In a way, this distrust of centralized governmental action echoes, at the political level, the complexification of economies and societies already mentioned; Mayntz (1995) thus argues that the increasing functional complexity of societies calls for new forms of societal and political governance.

Third, popular enthusiasm for LTSs has greatly declined. As a larger and larger part of the population gained access to LTS services, these services lost their 'wonderful' character. In advanced industrialized countries they are now taken for granted, as the current debate on 'universal services' demonstrates: the issue is now to ensure that virtually everybody has access to 'basic' services (such as telecommunications, postal services, electricity and water supply, etc.). At the same time, attitudes opposed to the expansion of LTSs are on the increase. Although not specific to these systems, the so-called 'NIMBY syndrome' ('not-in-my-backyard'; see Dear 1992) strongly affects the building of nuclear (and non-nuclear) power plants, new roads, airports or electrical grid lines. Also important in this respect is the increasing concern for the environment and for the well-being of 'future generations'. In particular, environmental externalities and the risk of catastrophic failures increase fears in a growing part of the population.

The traditional economic tools (cost–benefit analyses) seem inadequate to deal with these emerging regulatory issues. The demonstration of the economic, social and even environmental benefits of a project (assuming that a methodology can be agreed upon), are not appropriate responses to NIMBY attitudes; and cost–benefit analyses cannot easily integrate sustainable development concerns. Besides, as the debate on universal service reveals, the regulation of LTSs cannot be separated from the increasing concerns about the economic, social, and spatial fragmentation processes at work in contemporary societies (see Guy *et al.* in this volume); but the microeconomics of regulation and the macroeconomics of welfare are not easily reconciled.

Finally, the community of interests between LTS industries and national governments is also under increasing pressure. Corporate strategies now develop at other territorial levels, as the neologism 'glocalization' suggests: there is an increasingly global (supranational) competition for increasingly local (infranational) markets – mostly markets for urban infrastructures or services (see Lorrain 1995). Even state-owned companies are more and more at odds with their governments or with the legislator. Managers complain that their strategy is hindered by outdated legislation, or that governments are unable to provide their companies with new capital, due to fiscal crises, but that they are only too willing to appropriate their profits – for the same reason.

Taken together, these evolutions radically transform the frame of governance of LTSs, at the firm level (in terms of management), at the industry level (in terms of public oversight) and at the societal level (in terms of how LTSs shape, and are shaped by, societal governance patterns). They raise new, complex strategic and policy issues, some of which are the subject of the chapters in the second part of this volume.

Strategic and policy issues

So-called 'deregulation' is the most debated development in the LTS environment. Liberalization, the substitution of competitive markets for regulated monopolies, is a response to the perceived – and often real – inefficiency of regulated firms. Liberalization reforms enhance some forms of decentralized governance mechanisms within LTS industries. Technological change, for example, used to be undertaken within existing firms (see Tarr's study) or supported by governments (see Griset); it is now increasingly undertaken among competitive firms at an international level, beyond direct political influence. More generally, because it fosters the internationalization and the diversification of the activities of LTS firms, and because it is often accompanied by the privatization of previously public companies, liberalization reduces in various ways the influence of government intervention in LTS industries.

The chapters in Part 2 of this volume, by Summerton, Abbate, McGowan and Guy *et al.* all address the context of deregulation, and its implications for public policy and management in LTS industries. They provide useful complementary perspectives to mainstream economic analyses of regulatory reforms, which have been flourishing in the literature over the past decade.

Jane Summerton explores decentralized, non-hierarchical forms of management of energy systems at the regional (infranational) and the transnational levels. She contrasts traditionally dominant patterns of relations among firms (the territorial coupling of autonomous systems, or their full integration within a single system) with 'alliances', an intermediate pattern 'between coalitions and mergers' (Williamson 1990). Summerton argues that, although alliances (such as railway confederations or power pools) have always existed in LTS industries, the 1990s have seen the appearance of new forms of alliances, which she analyses as a means for 'recreating centres of managerial power' in the context of new corporate strategies induced by deregulation. Apart from economic motives, she argues that new alliances are also motivated by the need to bypass pressures, constraints or sensitive issues at a political/social level. She notes in conclusion the challenges posed by alliances to traditional local or national regulatory institutions.

In her study of standard design in the telecommunications industry (or more generally in information networks), Janet Abbate also notes the emergence of decentralized, non-hierarchical systems. The number of actors and of technologies within the industry is increasing rapidly, while the role played by public authorities or administrations is decreasing. As a result of these evolutions, designing and implementing telecommunication standards tends to be more and more difficult and the industry faces the risk of technical fragmentation. Based on a detailed and accurate account, Abbate sketches what the new role of public authorities in a decentralized industry could be. She argues that governments or administrative bodies should not try to 'assert authority where it no longer exists', but rather should use their power to coordinate standardization processes – for example by setting rules for standards development, accrediting standards organizations, and providing the industry with a body for standards testing, verification and information).

Francis McGowan's study points up another challenge to public regulation: the internationalization of LTSs. He argues that while many issues (monopoly power, standards requirements, environmental externalities and economic and social development concerns) still call for public regulation, the appropriate place for this regulation is increasingly at the international level, where scarcely any strong regulatory body exists yet. Since issues to be regulated often exceed the limits of specific LTS industries, and since firms tend to diversify into other activities, sectoral regulation is also under pressure, and McGowan makes a point for regional (such as the European Commission) or global (such as the World Trade Organization) regulatory institutions. These institutions, however, are as yet not fully effective. In particular, McGowan notes, 'the moves to tackle global environmental problems indicate the problems in securing the agreement of countries with very divergent interests'.

Simon Guy, Stephen Graham and Simon Marvin study the social, spatial and environmental implications of the 'profit logic' which is displayed by privatized firms in the liberalized UK utility industries. They uncover utilities' strategies of 'cherry picking' (i.e. being responsive to high-income, potentially profitable customers), 'social dumping' (i.e. trying to disengage from low-income ones), and demand-side management (with the aim of reducing investment in additional capacity). While some implications of these strategies are clearly beneficial to the public interest (in particular increased efficiency and environment-friendly demand-side management), others have been regarded as unacceptable. This was notably the case for the wave of water supply disconnections (a rather crude form of 'social dumping') that occurred in the immediate aftermath of privatization in that industry. The reaction of public opinion and the media was so strong as to compel the regulator to intervene. Guy, Graham and Marvin argue for increased regulatory intervention on social issues, although these issues are not explicitly part of the regulator's mandate.

An important lesson can be drawn from these four chapters – namely that so-called deregulation requires a lot of regulation. This could appear paradoxical, but it is not. Indeed, as the authors of these chapters strongly argue, the case for regulation has not miraculously vanished with liberalization; if anything it has got stronger. For example, emerging international diversified oligopolies may be more detrimental to the public than previous national (or local) industry-specific monopolies. Environmental and social concerns have not faded away either. The paradox is rather that, as the needs for regulation grow, governments seem keen to abandon most of their means of control over LTS activities. But there is more to the studies gathered here: they all carefully and accurately discuss what future forms of regulation could or should be. What should be the level, focus and modalities of regulation? How is regulation affected by the technological nature of LTS industries? How can the 'power plays' among actors involved in these industries be organized? By addressing these questions the studies gathered here go well beyond the conventional economic discussions of 'optimal regulation': the games of corporate actors (Summerton), the technical and institutional fragmentation in the telecommunications industry (Abbate), the geopolitics of regulation (McGowan) and the social, spatial and environmental dimensions of utility networks (Guy *et al.*), are crucial elements in the debate on what socially-effective regulation of LTSs might resemble.

The governability of large technical systems

In the existing literature, the many challenges to the traditional, hierarchical forms of governance of LTSs have been examined. Like many LTS scholars, I have implicitly argued that the weakening of hierarchies raises a number of serious issues, in particular regarding control of the negative external effects of LTSs. In the two convincing contributions in Part 3 of this volume, Alain Gras and Donald MacKenzie challenge commonly-held views on the scope of centralized control of complex technologies. MacKenzie argues that fully-centralized control is impossible, and Gras suggests that it is not even desirable.

In a theoretical study, MacKenzie examines the governability of a particular kind of LTS: he explores the conditions under which an agreement on comprehensive nuclear disarmament could be made permanent – in his words, how nuclear weapons could be 'uninvented'. Admittedly, uninventing systems is not the primary concern of most managers or regulators. However the validity of MacKenzie's insights goes well beyond this particular context. MacKenzie argues that, because of their systems dimension, the tacit knowledge (i.e. knowledge or expertise that cannot easily be formalized or made explicit) they include, and the actor-networks they are part of, a ban on nuclear weapons could be made permanent by the disappearance of the know-how or the skills (technological, organizational, institutional . . .) they rest upon. Encouragingly enough for its supporters, the fulfilment of this scenario would not require an unlikely centralized world government. By emphasizing in particular the role of tacit knowledge in technological development,

MacKenzie suggests that full centralized control of complex technologies is not possible: tacit knowledge is primarily local, and resists centralization. Complex technologies may require some degree of decentralized control, in the sense that some crucial actions may have to be implemented at a 'local' level. The role of tacit knowledge, however, is perhaps not so important in system design – as in MacKenzie's example – as it is in system operation (e.g. for air transport systems, nuclear power stations or power grid systems).

In another theoretically-oriented study, concerned with air transport 'large-scale technical systems', Gras adopts a different perspective. He argues that the traditional organization of these systems has reached its limits (in terms of capacity). In the current organization flight paths are centrally monitored by air traffic controllers. The ability of controllers to deal with only a limited number of 'flying objects' at one point in time is an important factor in the congestion of the system: it can be described as 'cognitive' – as opposed to 'technical' – congestion. One solution to the congestion problem is the full automation of traffic control; it is in line with the long-lasting process of centralization ('delocalization', in the words of Gras) in the management and control of many LTSs, a process which Gras examines in the first part of his chapter.[10] An alternative solution is considered: namely to devolve some scope for free initiative back to pilots. This would constitute a radical change for air transport systems; in particular it would challenge the strongly embedded preconception (among air transport engineers and many others) that the more centralized the control, the better the performance of the system. Gras convincingly argues that genuine 'decentralization' could offer an expedient way out of the congestion–complexification dilemma – not to mention the societal implications of such a change of perspective (see Gras 1997).

Hence both MacKenzie and Gras challenge the widely-accepted view that only centralized technological systems are governable. The next issue confronted by students of LTSs is then: how should the control problem be reformulated or transformed in order to dissociate (both conceptually and functionally) governability from centralized control? The concept of 'network', discussed in Part 4, may be one step in this endeavour.

Conceptualizing new governance models for LTSs

The concept of network recently gained considerable popularity in writings about LTSs, as the following assertion shows: 'The Internet is an internetwork of many networks [providing] a test bed for new developments in networking' (Quaterman 1990, quoted by Rochlin 1995). Social studies of LTSs also resorted to this concept to emphasize complex, heterogeneous, decentralized, self-organized, ungovernable, 'postmodern' or even 'transmodern' systems (Rochlin 1995). The three contributions in the final part of this volume help to capture the epistemological implications of the use of this fashionable concept.

It is as well to recall that the word has many different meanings. In his *Networks of Power*, Thomas Hughes (1983) uses the word in at least three different ways. First, as a synonym for infrastructure: 'a great network of power lines . . . is now superimposed on the industrial world' (Hughes 1983: 1); second, as a metaphor to describe entangled power relationships: 'I turned to an exploration of a broad range of factors, events, institutions, men, and women involved in complex networks of power – technical, economic, political, and social' (*ibid.*: x); and third, merely to emphasize links between various phenomena: 'I found that the network of evolving technologies often linked Germany with the United States because both were industrializing rapidly, and that England often provided a contrast to events and trends observed in the two other countries' (*ibid.*). None of these corresponds to the now dominant use of the word 'network' in LTS studies – which makes for a fourth meaning. This is what we shall now discuss.

Jean-Marc Offner's contribution to this volume provides an illuminating comparative history of the French notion of 'réseau' (network) and of corresponding technological systems. Offner distinguishes four meanings of the notion, each symbolized by a given type of LTS (although, Offner argues, the history of the notion and that of the real systems are partly independent). They are in chronological order: the 'circulatory machine' (epitomized by the railroads); the 'relational graph of customers' (electricity, telephone); the 'spatio-temporal information switching device' (high speed transport, electronic communication); and the 'decentralized coordinator' (conversational computers). As new meanings emerged, they did not replace, but rather combined with, previously existing ones to shape an increasingly complex notion.

The parallel with the systems perspective in technology is striking. In particular, Offner notes, 'the distinction made by Braun and Joerges (1994a) between first- and second-order systems seems . . . in line with our typology, with second-order systems corresponding to the fourth paradigm, namely the network as a decentralized coordinator, in which it is no longer a question of technical networks, but of networked organizations'. Indeed, Braun and Joerges explicitly relate second-order LTSs to the notion of network: 'the concept of second-order large technical system refers to the process of networking parts of different first-order systems for specific, macro-level social domains' (1994a: 26). In line with its various disciplinary origins (transaction cost economics, political science and sociology of organizations), the word 'network', as it is now used in LTS studies, emphasizes *inter-organizational transactions*.

This brings us back to the problem of control. In his contribution on the epistemological groundings of the systems approach, Reiner Grundmann emphasizes the heuristic value of the notion of network. Although he refers specifically to the Callon–Latour concept of actor-network, his remarks have a more general character: approaches using network-like metaphors (web, fabric, etc.), he argues, 'shift our attention towards shifting boundaries, contested terrains, manipulation of resources and of competitors, be they potential or real. *There is no longer one machine which can be optimized*' (original emphasis).

Bernward Joerges, in the concluding chapter, challenges the idea that the growing use of the notion of network reveals a change in the systems perspective on technology: 'the largest common denominator in the discussion on common characteristics [of LTSs] is probably the notion that LTSs might best be described as multi-actor systems – whether in the sense of groups assuming certain roles within organizations viewed as enveloping LTSs or in the sense of inter-organizational networks'. There are, however, at least two broad patterns of relations within a multi-actor system: one is where one dominant actor enrols other actors to serve his/her purposes; an altogether different pattern is where actors pursue interdependent but unidentical (possibly conflicting) purposes, and where no single actor is dominant. In the latter pattern, the outcome of the co-evolution of the system's actors is not pre-determined. Perhaps the phrase 'interorganizational networks' should be reserved for this pattern. Although a multi-actor approach is certainly a useful step in dealing with interorganizational networks, it is argued here that most LTS studies have privileged the first pattern of interactions within multi-actor systems.

Beyond this particular debate, Joerges provides us with a challenging exploration of LTS studies. Through his discussion of 'quasi-definitions and quasi-generalizations', 'metaphors', 'narratives', and 'stories not told', Joerges unveils many of the implicit assumptions in, and implications of, LTS discourses. He argues that the 'high variability discourses' which can be observed in LTS studies in fact result from the irreducible paradoxicality of the systems themselves, and thus should not be perceived as a weakness, but as a resource that should, if anything, be increased.

What's next on the research agenda?

All introductory essays of this kind seem bound (doomed?) to end with recommendations (or prophecies) about future research in the field. It is a bold endeavour to presume as to what one's colleagues should become interested in . . . and it also runs the risk of being useless. There is fortunately a way out of this trap: linking one's forecasts with already emerging research trends. This pragmatic approach basically assumes that one's colleagues have more than a vague idea of what they should be working on; it also benefits from (instead of fighting against) the significant momentum that characterizes trends in research interests. The more reasonable task for the 'introductory essayist' is then to emphasize what she or he views as the most promising areas for the problem under examination. In this respect, I shall briefly mention three useful trends in social studies of technology.

First, links between urban studies and LTS studies should be reinforced. Pioneer contributions to this area of research included those of Joel Tarr and Gabriel Dupuy, which analysed the rise of the 'networked city' (see e.g. Tarr and Dupuy 1988, and Dupuy 1991). It is now commonly acknowledged that studies of supranational systems (such as telecommunications, energy or air transport systems) and of urban technical networks (water supply and sewerage systems, for example) must be related (see Summerton 1992, Coutard 1995) – if only because of the unifying dynamics generated by regulatory reforms in all these industries. The relations between urban governance and infrastructural systems now deserve to be systematically investigated, in their economic, social, environmental and political dimensions. The French Research group on Networks (Groupement de recherche 'Réseaux') has been doing research in this area for a few years (see in particular Lorrain 1993; for a discussion on technical systems and territorial governance on a broader scale, see Barraqué 1997; for a general view of the territorial approach see Offner and Pumain 1996). A recent initiative along the same lines was taken by the Centre of Urban Technologies at Newcastle University (UK), in the form of a conference on 'Urban futures, technological futures'.[11]

Second, the conflictual dimension of multi-actor or multi-organization systems should be further investigated. The notions of governance and of network seem to implicitly convey the idea that it is all a matter of consensus building through interaction. Such, however, is not generally the case. Actors and organizations have conflicting interests, values and strategies. And inter-actor or inter-organizational transactions certainly do not always boil down to the manipulations of a machiavellian dominant actor in actor-networks. Conflicts are the rule, not the exception. What becomes of them? How are they managed? Can they be solved, and if so how? Conceptual and empirical materials are already available but could be further elaborated.

Third, as was frequently noted (recent instances include Summerton 1994b and Joerges in this volume), the systems approach to technological systems has thoroughly neglected users. This is obviously detrimental to a better understanding of the ins and outs of the governance of large technical systems; I shall therefore add my own voice to their call for accurate, in-depth studies of users and uses of these everyday technologies. It is essential to elucidate as far as possible the process of 'co-production' – by system-builders and users – of the purposes, performance, regulation and societal implications of large technical systems. Only then can the crucial issue of their (il)legitimacy be relevantly addressed.

Acknowledgements

I wish to thank Bernward Joerges and Jean-Marc Offner for their comments on an earlier version of this introduction.

Notes

1 'Personally, I am among those who hold that there is no clear connection between constructions of complexity in the engineering sciences, in social systems theory or in evolutionary biology. The same term covers different ideas and ideologies. In LTS research we find traces of all of them. But a term cannot become common coin in many disciplines without pointing to a great variety of subtle connections. In this sense, the considered use of this notion will help to achieve this rare good in scientific research: communication and cooperation across intellectual cultures' (Joerges 1996: 69). Doesn't the same go for governance as for complexity?

2 'Load factor is the ratio of the average load to the maximum load of a customer, group of customers, or the entire system during a specified period' (Hughes 1983: 218).

3 'An economic mix is an interconnection of power plants whose energy sources are utilized complementarily. . . . The operation of the plants would be carefully scheduled to take advantage of their complementary characteristics' (Hughes 1983: 367).

4 Cf. the third international conference on the Development of LTSs ('Trans-Border Aspects of LTSs), held in Sydney in 1991.

5 A well-known example is the integration of direct and alternating currents in Westinghouse's 'universal power system' (see Hughes 1983).

6 When a similar transformation occurred in Britain, it was described by a British observer as 'perhaps the greatest peacetime operation in this nation's history' (quoted by Tarr).

7 But not all of them; see in particular Summerton 1992.

8 Not the least in terms of the time periods considered: the late nineteenth century for Griset and Chatzis, the second half of the twentieth century for Tarr and Kaijser).

9 Early LTS development has even met with popular enthusiasm. Railways, the telegraph, the telephone, electricity, the car or air transport have all been perceived as wonderful (as well as fearful) phenomena. Beyond this first perception of LTSs as 'magic', it is interesting to note that the uses which made LTSs so popular were seldom part of the initial design. For example, the telephone was first conceived as a means for broadcasting music scores, theatre plays or other information; its use as a conversational device was invented by early users (Fischer 1992). Electricity was used for decades as a means for lighting public places before it entered homes. Castoriadis (1992) has noted the ambiguous nature of social attitudes towards technologies, often combining a fascination for technical artefacts (or systems) with a pessimistic view of the societal implications of technology. Flichy (1995) analyses the complex interactions between technical prototypes and social expectations ('imaginaire social') in the innovation process. The success (or the failure) of an innovation, he argues, results from a complex process of stabilization among technical components of the artefact and among actors engaged in technical activities.

10 Gras does not argue that all LTSs have become more and more centrally managed and controlled over time. Rather he suggests that a number of LTSs (e.g. energy supply, railways or air transport systems) are part of (i.e. both supporting and

supported by) a broader process of empowerment of 'technoscience' in modern societal governance. He calls these particular LTSs large-scale technical systems (in French, 'macro-systèmes techniques') to distinguish them from other LTSs.

11 Conference 'Urban futures, technological futures', Durham (UK), 23–25 April 1998. See also Graham and Marvin (1996) for a remarkable achievement in bringing together the two perspectives.

References

Barraqué, Bernard (1997) 'Gouverner en réseau en France: les agences de l'eau', in Michel Gariépy et Michel Marié (eds) *Ces réseaux qui nous gouvernent?*, Paris: L'Harmattan (Coll. Villes et Entreprises).

Braun, Ingo and Joerges, Bernward (1994a) 'How to recombine large technical systems: the case of European organ transplantation', in Jane Summerton (ed.) *Changing Large Technical Systems*, Boulder (CO): Westview.

—— (eds) (1994b) *Technik ohne Grenzen*, Frankfurt am Main: Suhrkamp.

Castoriadis, Cornelius (1992) 'Technique', in *Encyclopaedia Universalis* (Vol. 22).

Coutard, Olivier (1995) 'The Paris conference on "Territorial Technologies"', *Flux, Cahiers scientifiques internationaux Réseaux et Territoires* 21: 5–9.

Dear, Michael (1992) 'Understanding and overcoming the NIMBY syndrome', *Journal of the American Planning Association* 58, 3.

Dupuy, Gabriel (1991) *L'Urbanisme des réseaux, théories et méthodes*, Paris: Armand Colin.

Fischer, Claude (1992) *America Calling: A Social History of the Telephone to 1940*, Berkeley (CA): University of California Press.

Flichy, Patrice (1995) *L'innovation technique. Récents développements en France vers une nouvelle théorie de l'innovation*, Paris: La Découverte.

Graham, S. and Marvin, S. (1996) *Telecommunications and the City: Electronic Spaces, Urban Places*, London: Routledge.

Gras, Alain (1997) *Les macro-systèmes techniques*, Paris: Presses universitaire de France (Coll. Que sais-je?).

Hughes, Thomas P. (1983) *Networks of Power. Electrification in Western Society 1880–1930*, Baltimore (MD): Johns Hopkins University Press.

—— (1987) 'The evolution of large technological systems', in Wiebe Bijker, Thomas Hughes and Trevor Pinch (eds) *The Social Construction of Technological Systems: New Directions in the History and Sociology of Technology*, Cambridge (MA): MIT Press.

—— (1989) *American Genesis. A Century of Invention and Technological Enthusiasm*, New York: Penguin.

—— (1994) 'Technological momentum', in Merritt Roe Smith and Leo Marx (eds) *Does Technology Drive History? The Dilemma of Technological Determinism*, Cambridge (MA): MIT Press.

—— (1996) 'Fifteen years of social and historical research on large technical systems' (interview by Olivier Coutard), *Flux, cahiers scientifiques internationaux Réseaux et Territoires* 25 (July–September): 44–7.

Joerges, Bernward (1988) 'Large technical systems: concepts and issues', in Renate Mayntz and Thomas Hughes (eds) *The Development of Large Technical Systems*, Frankfurt am Main: Campus.

—— (1996) 'Large technical systems and the discourse of complexity', in Lars Inglestam (ed.) *Complex Technical Systems*, Stockholm: Swedish Council for Planning and Coordination of Research (Affärs Litteratur), 55–72.

Jouve, Bernard, Lefèvre, Christian and Offner, Jean-Marc (1995) 'La gouvernance urbaine', *mimeo*, Marne-la-Vallée: ENPC-Latts (April).

La Porte, Todd (ed.) (1991) *Social Responses to Large Technical Systems: Control or Anticipation*, Dordrecht (NL): Kluwer.

Lorrain, Dominique (1993) 'Les services urbains, le marché et le politique', in Claude Martinand (ed.) *Financement privé des équipements publics*, Paris: Economica.

—— (1995) 'La grande entreprise urbaine et l'action publique', *Sociologie du travail* 37, 2: 199–220.

Marin, Bernd and Mayntz, Renate (eds) (1991) *Policy Networks: Empirical Evidence and Theoretical Considerations*, Frankfurt am Main: Campus, and Boulder (CO): Westview.

Mayntz, Renate (1995) 'Progrès technique, changement dans la société et développement des grands systèmes techniques', *Flux, cahiers scientifiques internationaux Réseaux et Territoires* 22 (October–December): 11–16.

Mayntz, Renate and Hughes, Thomas P. (eds) (1988) *The Development of Large Technical Systems*, Frankfurt am Main: Campus.

Offner, Jean-Marc and Pumain, Denise (eds) (1996), *Réseaux et Territoires, significations croisées*, La Tour d'Aigues (F): Editions de l'Aube.

Quaterman, John S. (1990) *The Matrix: Computer Networks and Conferencing Systems*, Burlington (MA): Digital Press.

Rochlin, Gene (1995) 'Pris dans la toile: réseaux, mutations et conformité à l'ère de l'informatique', *Flux, cahiers scientifiques internationaux Réseaux et Territoires* 22 (October–December): 17–29.

Salsbury, Stephen (1988) 'The emergence of an early large-scale technical system: the American railroad network', in Renate Mayntz and Thomas P. Hughes (eds) *The Development of Large Technical Systems*, Frankfurt am Main: Campus.

Summerton, Jane (1992) *District Heating Comes to Town: The Social Shaping of an Energy System*, Linköping (Sw): Linköping university studies in arts and science.

—— (ed.) (1994a) *Changing Large Technical Systems*, Boulder (CO): Westview.

—— (1994b) 'The systems approach to technological change', in Jane Summerton (ed.) *Changing Large Technical Systems*, Boulder (CO): Westview.

Tarr, Joel and Dupuy, Gabriel (1988) *Technology and the Rise of the Networked City in Europe and America*, Philadelphia (PA): Temple University Press.

Thompson, Grahame, Frances, Jennifer, Levacic, Rosalind and Mitchell, Jeremy (1991) *Markets, Hierarchies and Networks. The Coordination of Social Life*, London: Sage.

Veltz, Pierre (1996) *Mondialisation, villes et territoires. L'économie d'archipel*, Paris: Presses universitaires de France (Coll. Economie en liberté).

Williamson, Olivier E. (1990) 'The firm as a nexus of treaties: an introduction', in M. Aoki, B. Gustafsson and O.E. Williamson (eds) *The Firm as a Nexus of Treaties*, London: Sage.

Part 1

ON TRADITIONAL MODES OF GOVERNANCE

2

TRANSFORMING AN ENERGY SYSTEM

The evolution of the manufactured gas industry and the transition to natural gas in the United States (1807–1954)

Joel A. Tarr

It has been neatly said that the self-styled cities become such in reality when they are lighted with gas.

The History of Clinton County, Iowa (1879)

Introduction

This chapter focuses on the development of the manufactured gas industry in the United States and the transition to natural gas, identifying key innovations and turning points in its technological and policy history. The manufactured gas industry was composed of a large number of urban utilities, each of which had its own centralized production center and piped distribution system. That is, the industry consisted of spatially separated and limited, but similar technical systems for which there were only very restricted technological means of product interchange.[1] These systems varied greatly in size and resembled other urban networked systems such as water supply and sewerage (Tarr and Dupuy 1988).

Because gas companies used the public streets for gas delivery, they required public franchises and were subject to municipal and later state regulation. The industry was capital intensive and used an imbedded technology, imparting to it the problems of asset specific investment and requiring firms to seek franchises that were long rather than short in duration. These system characteristics gave gas utilities natural monopoly features and their pricing was often immune to market forces, exposing them to political criticism. However, while some municipally-owned gas works existed, and major political battles were fought over rate and ownership issues, the industry remained largely private. As in the case of electrical power, the industry came to favor state over local regulation because of the uncertainties of local circumstances and the behavior of local politicians. That is, state regulation was acceptable to utilities if it was 'the only way to forestall municipal control' (Troesken 1994: 113–15).

In the twentieth century, critical innovations occurred in the development of long-line pipelines, and utilities increasingly converted to natural gas mixed with manufactured gas or natural gas alone. Eventually, most local firms ceased as production units and focused only on natural gas distribution. They thus became spatially connected technical systems and part of much larger, nationwide networks that linked markets with production centers over long distances, often crossing state boundaries. As such, natural gas pipeline companies became subject to federal regulation. The manufactured gas industry, as it had existed for almost a century and a half, died-out almost completely, although part of its skeleton – its local distribution systems – remained, with a transfusion of a different substance in its arteries.[2]

Coal gas domination, 1816–75

The early years in the development of the (manufactured) gas industry were marked by important innovations in regard to fuel, distillation technology, and distribution. Initially, US manufactured gas firms used technology developed in Great Britain. Entrepreneurs founded the first public gas utility in Baltimore in 1816; it had a central gas plant from which gas was distributed to users via a networked system of mains (King 1950). The form of this system was modeled after that used for water works (hence, 'gas works'), as were other urban networked systems (Elton 1958: 270–2, Benjamin 1895: 827–9). Other important system components included gas holders, or 'gasometers' for storage, distribution systems, and meters.

Wooden pipes had originally been used for distribution, but in 1820 or so builders began substituting British-made cast iron pipes for the original wooden pipes; by the 1830s, American foundries were producing their own cast iron pipe (Elton 1958: 271–2, Ershkowitz n.d.: 31–8, Williamson and Daum 1959: 40). Gas was initially used primarily for street illumination, but cost reductions plus innovations such as improved burners and fixtures and metering led to a broader domestic, commercial, and industrial lighting market. Gas works were an important element in the networked city that developed in the middle and late nineteenth century (Tarr and Dupuy 1988). Dissemination among the urban network, however, initially proceeded slowly, probably because of a scarcity of engineering know-how, technological and fuel uncertainties, and capital shortages.

Any organic material can be used as a feed stock for the production of gas, but with various degrees of efficiency. The Baltimore Gas-Light Company, for instance, the first urban firm (1816), initially used pine tar as the feedstock for its gas production. Other cities used rosin, British coal, wood, and distillate of turpentine. Utility managers originally avoided coal because of its impurities, but British gas manufacturers developed methods of removing them around 1815–20, and the technique was transferred to the United States in 1822 (King 1950). Bituminous coal, however, was not easily accessible, and fuel costs kept prices relatively high, limiting possibilities of enlarging the market for improved lighting. While high quality bituminous coal was available in Western Pennsylvania, the Allegheny Mountain barrier confined its use to the Pittsburgh area. It was not until the late 1840s and early 1850s, when the Monongahela and Youghiogheny rivers were slack watered and rail connections were established from Philadelphia to Pittsburgh, that high quality Western Pennsylvania bituminous 'gas coal' became available to eastern gas works (Binder 1974: 29–39, Commonwealth of Pennsylvania 1875: 88–97, Eavenson 1942: 188–202).

Lower fuel costs that reduced the cost of light and increased demand resulted in a large increase in the number of city gas works founded in the 1850s and 1860s. US Census statistics (admittedly incomplete) for the nineteenth-century gas industry report that between 1850 and 1860 the number of manufactured gas establishments jumped from 30 to 221, or an increase of over 600 per cent – a much greater rise than that for urban growth. By 1870, the total reported was 390 (US Bureau of the Census 1889: 699). Coal gas manufacture also provided the potential for firms to recover and sell byproducts such as coke, tar, light oil, and ammonia. However, unlike Great Britain, the US possessed no market for such products (aside from the coke) before the late-nineteenth century. The nation lacked a coal-based chemical industry and was almost completely dependent upon imports for dyestuffs, drugs, explosives and other products of coal tar and light oil origin (Rhodes 1966: 3–14).[3]

By the 1870s, there were over 400 gas manufacturing firms, located in all regions of the country (US Bureau of the Census 1870: 439). Price, however, limited home gas service to a relatively affluent segment of the potential market. Gas companies had originally hoped to make large profits from street lights, but municipal franchises applied stringent regulations as to the numbers of public street lights and rates paid, and many firms defaulted on this part of the contract or were continually petitioning city councils for rate increases (King 1950: 38–9, Moehring 1981: 99–109). Because costs were high and margins low, gas companies were reluctant to make the investment necessary to expand to potentially lucrative but risky lower income markets. This conservatism duplicated a pattern characteristic of the early private water companies, but unlike water supply, widespread public ownership did not result because of competitive pressures (Jacobson and Tarr 1995: 10–14). The industry was at risk from lower priced fuels and energy sources, and these gradually appeared, beginning with coal oil in the 1850s, followed by kerosene and electricity in the postwar period (Williamson and Daum 1959: 40).

Challenges from competing fuels, 1875–1927

Lighting

This period saw the development of substantial competition for the urban lighting markets, important industry adaptations, and new regulatory pressures. The first competitor for the lighting market was the kerosene lamp, an outcome of the discovery of oil in Pennsylvania in 1859. The kerosene lamp was especially important in capturing the low end of the illumination market. Improvements in petroleum refining and transportation techniques and the discovery of new sources of supply, resulted in reduced oil prices. By the 1870s and 1880s, most of the urban working class, as well as a substantial part of the middle class, used kerosene lamps for domestic illumination (Williamson and Daum 1959: 55–60).

Competition from electricity began with arc lamps in the 1880s, when electrical companies made inroads into the market for street illumination. By the 1880s, however, many gas companies had found these lighting contracts unprofitable, often because of franchise limitations. Domestic lighting accounted for 90 per cent of gas revenues, and arc lamps posed no threat in this market. The incandescent lamp, which appeared in 1881, directly competed with interior gas light. Thomas Edison, inventor of the incandescent lamp and builder of the first central electric generating station, had studied the gas industry carefully, searching for an 'electrical analogue to the gas system'. Competition from both the kerosene lamp and the incandescent electric light 'retarded the growth of the gas industry', forcing it to change its production methods and eventually to develop new markets (Passer 1953: 195–200).

Carbureted water gas provided the first major innovation designed to meet this challenge. The concept behind water gas, which was first observed in 1780 and patented in France in 1834, involved the action of steam upon incandescent carbon, usually anthracite coal or coke. When the steam, composed of oxygen and hydrogen, came into contact with the incandescent carbon, it produced a fuel gas composed of hydrogen and carbon monoxide, as well as other products. The gas had a heating value of about 300 Btu (British thermal units) per cubic foot, burned with a blue flame, and had little illuminating power. But because the market for manufactured gas before the late-nineteenth century was almost entirely in lighting, water gas was not competitive.

In 1875, however, Thaddeous S.C. Lowe, a Pennsylvania inventor, patented a process to enrich blue gas by spraying it with liquid hydrocarbons and thermally cracking the gas to form carbureted water gas and tars (Block 1966). The United Gas Improvement Company (UGI) of Philadelphia, formed for this purpose, acquired the Lowe Process in 1882. UGI aggressively pushed the technology, overcoming opposition to water gas. Because of the rapid expansion of the petroleum industry, oil to enrich the gas was available at a relatively cheap price. The US gas industry proceeded to shift much of its manufacturing processes from coal to carbureted water gas (US Bureau of the Census 1895: 705, 1900: 711). Great Britain (first water gas plant 1890) and other European nations, which did not have access to inexpensive petroleum at that time, did not adopt carbureted water gas to the same extent as did the United States. The US gas industry, therefore, became increasingly dependent on oil, while the European industry remained largely coal based (Matthews 1987: 978).

Carbureted water gas had a number of advantages over coal gas. Its production required less space, capital, and labor than coal gas, as well as smaller storage capacity, because plants could be started up in a relatively short time. Many coal gas plants installed auxiliary carbureted water gas machines to increase their flexibility in regard to load, mixing the coal and water gas. Without the auxiliary water gas equipment, firms would have had to add storage or construct additional retorts, which would have been idle for many hours in the day. Carbureted water gas also had a higher illuminating value and a flexible heating value; a plant could produce gas varying in heating value from 300–800 Btu per cubic foot. Utility operators could produce carbureted water gas with widely varying Btu content by changing the amount of oil cracked into the blue gas. Higher Btu carbureted water gas could be mixed with gas with a lower Btu grade from blue gas or coal gas retorts, thereby increasing the available gas volume without the necessity of increasing the number of water gas sets (Hasiam and Russell 1926: 618–25, Parr 1932: 148, Passer 1953: 195–6).

The addition of petroleum hydrocarbons to water gas provided a gas that had illumination values equal to coal gas but at a lower cost. Most municipal franchises to gas companies required illuminating power of between 15–17 candlepower. The gas burners most commonly adopted before the invention of the Welsbach mantle in 1886 used the luminosity of the flame to fulfill legal candlepower requirements. Therefore, when carbureted water gas was burned as a fuel, the burners produced an equal or better flame at a lower cost. The lamps, however, like coal gas, depended upon the incandescence of the carbon particles in the flame for their illuminating power (Collins 1934: 205–10).

In 1885, Carl Auer von Welsbach of Germany, working in the laboratories of Robert Wilhelm von Bunsen, produced another major gas innovation. Welsbach patented a gas mantle made of cotton fabric impregnated with rare earth. When the mantle was placed near a gas flame from a Bunsen burner, it 'became incandescent' and produced a steady white light of 30–50 candlepower. This increased the light that could be obtained from a given quantity of gas about six times (Clark 1916: 666, Passer 1953: 196–8, Schivelbusch 1988: 48).

At the time of the introduction of the Welsbach mantle, municipal franchises usually stipulated a candlepower standard for gas ranging from about 14–16 for coal gas and 18–36 for water gas (Blake 1922: 1009). The illuminating power of the Welsbach mantle, which was not dependent on a higher candlepower gas rating, made these standards obsolete. Beginning with the Wisconsin Public Utilities Commission in 1908, states and municipalities gradually began requiring that gas be measured in Btu or heat units per cubic foot of gas rather than lighting units (Tussing and Barlow 1984: 16).

Other energetic uses of gas

In spite of the success of the Welsbach mantle, competition from electricity consistently ate into the market share of the gas industry. The use of gas solely for lighting resulted in poor load factors for gasworks. In order to meet competition and also to use their technology more efficiently by load spreading, utilities sought to enter new markets. Electrical utilities had followed such a strategy, attempting to attract streetcar firms as customers and selling appliances – now the gas companies did the same. The gas industry moved increasingly to markets outside of lighting, such as cooking stoves, water heating, and house heating (Kaijser 1993: 195–207, Rose 1995: 28–38). The shift away from light and into household appliances reflected industry recognition that manufactured gas was a fuel or energy source. The strategy was also a 'technical mimicry' of electricity – an attempt by an older technology to survive by copying a more recent technology (Platt 1991: 139–97, Passer 1953: 196–200, Schivelbusch 1988: 48–9, Tussing and Barlow 1984: 16).

Even though the gas industry developed substantial market share in the home utility domain, it still faced excess productive capacity and the need to develop new markets. About 1900 or so, gas firms moved into the industrial sector. Efforts were made to market gas engines, as well as to sell gas for heating purposes. Potential industrial customers, however, often found it more economical to use low Btu gas to meet their energy requirements rather than high Btu carbureted water gas. Frequently they used producer gas, a low-grade but inexpensive gas made directly on site from coal (Steere 1922: 91–4). It was also difficult for gas companies to enter industrial markets because of state requirements for a fixed number of Btu per cubic foot (usually 450–550 Btu) in their gas, which raised gas prices. This Btu rating was higher than many industries desired or would pay for, limiting the number of industrial customers.

The gas industry, therefore, began to push for the sale of gas using Btu as a standard and sold by the *therm* (100,000 Btu) as was done in Great Britain, rather than by the cubic foot. Industries could then obtain less costly low Btu gas to supply the heat they needed in their processes (*Gas Age* 1921: 216, Parr 1932: 149, Weber 1922: 1095–106). Changes in this regard were reported in the 1920 Census of Manufactures. The census reported that the prevailing

Btu standard was lower in 1919 than it had been in 1914, as the number of gas firms providing high Btu gas sharply dropped. Between 1919 and 1927, industrial use of manufactured gas increased from 70.4 billion to 136.4 billion cubic feet (US Bureau of the Census 1930: 719–20, Leinroth 1928: 1211–17).

Gas utilities also moved into the area of electricity supply, investing in generating equipment or acquiring existing electrical companies. In 1887, about forty gas companies supplied electric lighting, usually arc lights; by 1889, the number was 266, or about 25 per cent of US gas companies; and in 1899, 362 or nearly 40 per cent of gas companies provided electric lighting. A counter trend was for electrical utilities to acquire gas utilities, such as was done in the case of Pennsylvania Power & Light Company, Philadelphia Electric, or the Interstate Power Company of the Middle West, especially after the First World War (Passer 1953: 199–200, Wainwright 1961: 213–16).

Byproducts

Water gas possessed an important disadvantage compared to coal gas in terms of potential profits – the type of byproducts produced. Markets had developed in the late-nineteenth century for coal gas byproducts such as tar used for creosote in railway tie preservation, pitch for roofing and waterproofing, refined tar for road surfacing, and ammonia, cyanides or phenolic compounds for chemical industries. The carbureting process did produce byproducts, but they were not as salable as those from coal gas.

Between 1914 and 1929, the amount of manufactured gas distributed expanded from approximately 204 billion cubic feet to over 450 billion cubic feet. A little more than 50 per cent of the gas produced in 1929 was water gas, about 40 per cent was classified as coal gas, although the sum also included gas from byproduct ovens, and 6.2 per cent was oil gas. Between approximately 1916 and 1929, the steel industry shifted from the beehive coke oven to the byproduct oven. The byproduct coke oven captured the valuable coal chemicals previously lost, and also produced a gas surplus. Byproduct ovens could be heated by 35 to 40 per cent of the coal gas produced in the oven, leaving approximately 60 per cent to be sold. Coke-oven gas had a heating value of about 560 Btu and could be easily sold to other customers (Harkins *et al.* 1984: 17–22, Hasiam and Russell 1926: 648–51). By 1932, byproduct coke firms produced 25 per cent of the manufactured gas distributed to domestic, commercial, and industrial customers in the United States (Tyron 1928: 333–4).

During the 1920s, the manufactured gas utilities themselves moved into the byproduct industry. In order to meet peak period requirements, utilities with carbureted water gas technology had to manufacture large amounts of coke. By utilizing byproduct ovens rather than retorts, utilities could combine gas manufacture with coke manufacture to meet peak load requirements and produce a higher quality of coke for sale than that produced in retorts. Entrepreneurs also built a number of merchant plants that had as their chief products foundry coke, domestic coke, and coke for making water gas, as well as producing gas to sell under contract to public utilities (Tyron 1928: 333–4). By the end of the decade a clear trend existed for manufactured gas utilities to adopt carbonization equipment that produced both high quality gas and high quality coke; by 1932, 18.7 per cent of the manufactured gas produced by gas utilities was coke oven gas. Thus, as one text on fuels noted in 1926, 'the gas and the coke industries gradually are merging as far as equipment requirements are concerned, and . . . the tendency is toward the same type of plant for both branches' (Hasiam and Russell 1926: 651, Anderson and DeLawyer 1995: 83–99).[4] These developments are shown in Figure 2.1.

The transition to natural gas, 1927–54

The development of natural gas and its regulatory implications

Natural gas is a substance that occurs in nature under three conditions: in porous rock that forms a gas field (nonassociated gas), in association with petroleum deposits (casinghead or associated gas), and mixed with the oil so that it must be separated from it. Usually there is no oil field that occurs without gas and it is the pressure provided by the gas that is responsible for bringing the oil to the surface. Both natural gas and manufactured gas are composed of hydrocarbons – but they differ in heating value as well as other characteristics (*Gas Engineers Handbook* 1977: 2/17, 2/19). Manufactured gas usually averages between 550–600 Btu, while natural gas is about 1,000 Btu. Natural gas is not distributed evenly throughout the environment. It was first discovered in Fredonia, New York in the 1820s, and the Fredonia Gas Light Company, the nation's first natural gas company, was formed in 1858. Early discoveries were also made in West Virginia and Western Pennsylvania. When petroleum was discovered in Western Pennsylvania in 1859, natural gas came with it as a useless byproduct. Gas from these wells, as well as from other oil fields in Illinois, Indiana, and the southwest, was initially vented or flared because of the absence of markets within transportable distance. In 1872, the nation's first natural gas pipeline (2 inches in diameter, cast-iron) was used to transport gas five miles into Titusville, Pennsylvania (Stotz and Jamieson 1938: 67–93, Thoenen 1964: 103–9).

Pipeline technology, however, was limited, and leakage was high. The inability of the industry to produce long pipelines, therefore, constituted a 'reverse salient' (Hughes 1983), restricting the ability of producers to get natural gas to distant markets. Only markets within a relatively short distance of natural gas fields could benefit from the fuel, using it for domestic and

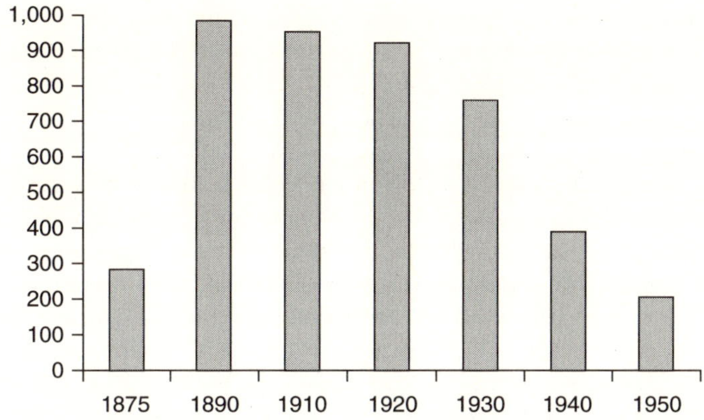

Figure 2.1 US manufactured gas plants, 1875–1950

Sources: Saward, Frederick E. (1878) *American Gas Making*, New York: Bedell and Bro; Eng, Robert (1985) *Survey of Town Gas and Byproduct Production and Location in the US, 1880–1950*, McLean (VA): Radian Corp (unpublished report).

Note The data supplied in this chart underestimate the number of gas manufacturing installations in the US during the 1890–1950 period. It only counts municipal manufactured gas plants listed in *Brown's Dictionary of North American Gas Plants*, and *Brown's* provides an underestimation. In addition, there were hundreds of small gas plants at industries and institutions. For some corrections to the data, see Hatheway, Allen W. (1997) 'Estimated number of manufactured gas and other coal-tar sites in the US', *Environmental and Engineering Geoscience* 3: 141–2.

industrial purposes. In the late-nineteenth and early-twentieth centuries, there were four regions where natural gas was piped into cities for use: Western Pennsylvania and West Virginia, northern and central Indiana, locations around Los Angeles including the San Joaquin Valley, and eastern Kansas. The fields that these urban markets drew from, however, were quite shallow, and by the early twentieth century they were severely depleted.

In the first decades of the twentieth century, additional natural gas discoveries were made in the Southwest, including the huge Panhandle, Hugoton and Monroe fields. Gas from these fields was initially wasted because of the absence of nearby markets. Technology to produce long pipelines did not exist, and before 1920 seamless-tube mills seldom produced pipe over 6 inches in diameter or longer than 25 feet; the longest pipelines were just under 200 miles. In the 1920s, however, significant breakthroughs occurred in welded pipe technology, making possible the construction of long-distance pipelines (Du Puy 1930: 55–8, Davis 1935: 3–4, Castaneda 1993: 15–18, McGannon 1964: 826–59). These developments were accompanied by improvements in field methods of ditch-digging and the use of protective coatings. In addition, the mathematical understanding of the role of tensile steel strength, pipe thickness and diameter, pipeline capacity, gas pressure and compressor station spacing to pipeline capacity, greatly increased. Pipeline construction had passed beyond the craft stage into one where decisions were made by mathematical analysis, and the reverse salient had crumbled (Davis 1935: 4).

These improvements resulted in the construction from 1925 to 1935 of a number of pipelines ranging from 200 to over 1,000 miles in length and connecting southwestern and California gas fields to urban markets (Tussing and Barlow 1984: 34–7). After the depression and the war, which slowed down pipeline construction, the middle Atlantic States and New England were the last major regions to be linked by pipeline with natural gas fields. New York City and Philadelphia were considered especially prized markets, with New York City alone consuming 40 per cent of the nation's manufactured gas. These cities were supplied with natural gas in 1947 (Tussing and Barlow 1984: 45–51, Castaneda 1993: 16–65, Castaneda and Pratt 1993, Gilmer 1981: 129–32).

Since utility regulation existed entirely on the municipal and state levels, the interstate gas firms were essentially unregulated. Concern over unethical and corrupt accounting and financial practices by utilities in general led to congressionally-ordered Federal Trade Commission (FTC) investigations into the industry. These investigations lasted from 1928 to 1935, and the resulting FTC report revealed an industry characterized by excessive waste of natural gas, pyramiding, stock watering, overvaluation, and a lack of competition. Four regionally based holding companies controlled approximately 60 per cent of natural gas production and the associated pipeline mileage (Castaneda 1993: 16–65, Sanders 1981: 24–35). The FTC concluded that, since the natural gas industry was unique in that the gas had to be used as soon as produced, 'the public concern is that it be produced only as needed, but that it be produced and delivered when needed, and then that it be used to the highest economic and social advantage' (Sanders 1981: 34).

Congress's response to these findings was to pass several legislative acts concerning public utilities. The Public Utility Holding Company Act of 1935 forbade large gas distribution companies from owning and controlling both production and pipeline firms. As a result, between 1935 and 1947, holding companies divested themselves of 306 utility companies including 113 gas companies. Congress then passed the Natural Gas Act of 1938. This act assigned the Federal Power Commission, which the Federal Power Act of 1935 had given the power to regulate interstate electric power transmission, the power to control many aspects of the operations of natural gas companies including rate setting. In addition, the Act attempted to control the problems of erratic supply and waste by only certifying a pipeline company to operate in interstate commerce if it had an adequate long-term supply of natural gas (*c.* twenty years) to serve its intended markets (Sanders 1981: 42–58, Castaneda 1993: 28–33).

The conversion process

Conversion to natural gas was a formidable undertaking. However, while considerable scholarly attention has been paid to the question of natural gas regulation, the logistics of the actual conversion of manufactured gas distribution systems into natural gas distribution systems has been largely overlooked.[5] Conversion, it should be noted, was not always seen as desirable by manufactured gas companies, even though there seemed to be many advantages. Some utility managers worried that natural gas supplies would prove insufficient to maintain their service commitments. Corporations were also reluctant to abandon their substantial investments in gas manufacturing plant and equipment. In 1926, for instance, the California Railroad Commission ordered resistant Los Angeles Gas and Electric Company and Southern California Gas to convert their entire systems from oil-based water gas technology to natural gas, arguing that by so-doing, customers would benefit from lower prices. On the other hand, in 1936, concerned about supply depletion and interruptions, the Michigan Public Utilities Commission denied the Grand Rapids Gas Light Company the right to shift from manufactured to natural gas, insisting that a 50/50 mixture of natural and manufactured gas be used (Littlefield and Thorne 1990: 81–4, *Gas Age* 1937: 86).

Conversion to natural gas as a fuel did not require a new municipal distribution system since the gas could be circulated in the same pipes and consumed by the same appliances used by manufactured gas. However, the different Btu ratings of the gases required that these appliances be adjusted, and natural gas odorized to prevent asphyxiation. Engineers and managers had to decide whether to use natural gas or natural gas mixed with manufactured gas, a decision often reached by experimentation. However, as was noted in the previous paragraph, regulatory commissions sometimes overturned this decision. Among the cities that chose mixed gas before the war were Chicago (1931) and Minneapolis (1935), which converted to 800 Btu mixed gas, New Orleans (1928), which chose a mixed gas of 600 Btu, and, Evansville, Indiana (1932), which used 800 Btu mixed gas for the transition period (*Gas Age* 1937: 86, Brown 1931: 161–2).

The 1946 Federal Natural Gas Investigating Committee recommended that natural gas be used primarily for enrichment purposes. This strategy would expand the production capacity of existing plants, provide backup in case of supply interruptions, and furnish peak load shaving abilities. A number of firms followed the recommendation for mixed gas (Grozier 1953: 39–41). In 1945, for instance, 301.4 million therms of manufactured gas and 988.5 million therms of natural gas were used to produce mixed gas; by 1954, this total had reached 936 million therms of manufactured gas and 2,012.7 million therms of natural gas (AGA 1956: 61). Other utilities, especially after the Second World War, went directly to natural gas, usually to avoid the costly necessity of adjusting gas appliances twice. Among those cities that went directly to natural gas were Davenport, Iowa (1934), Washington, DC (1946), the New York City borough of Staten Island (1949), and New York City's Consolidated Edison (1951). In many cases even utilities that had shifted completely to natural gas kept their gas making equipment for emergencies and for peak shaving purposes.

Once the decision concerning type of gas to be used had been made, utilities faced major organizational tasks of conversion and adjustment, regardless of whether mixed or natural gas had been chosen. In 1931, an American Gas Association Subcommittee issued a report on 'The study of factors involved in change-over from manufactured to natural gas'. The committee made four major recommendations: (1) that cities be divided into districts for conversion purposes and that each district be converted to natural gas separately; (2) that those handling the conversion task be thoroughly trained and that the work force be large enough that the largest district could be covered in a week; (3) that customers be kept fully informed at all times of the conversion process; and (4) that a universal adapter be used to adjust orifices. The committee also warned of the drying-out effects of the natural gas on joints and valves in a system that had formerly only carried 'wet' manufactured gas, with resultant leaks. The committee concluded, however, that compared to manufactured gas, natural gas presented fewer hazards (Wehrle 1931: 155–6). In addition to this committee report, utilities communicated extensively with each other about the conversion processes and the pitfalls to be avoided. Before converting to natural gas, utilities would send delegations to visit and observe the experience of cities undergoing conversion. Learning thus became a cumulative and industry-wide process (see for example Abell 1929: 36–71, Anderson 1936: 703–4, *Business Week* 1951: 108–13).

The largest single task facing a utility converting to natural gas was adjusting home appliances to the new gas – a task necessitated by natural gas's Btu rating of about 1,000 and the average 550 Btu rating of manufactured gas. Management of conversion was a major organizational task that could involve thousands of specially trained workers (Begeebing 1931: 60–1, *Gas Age* 1947: 26–8; see for examples Plata and Ferrarini 1949: 8–11, Johnson 1949: 38–43, 62–70 and Grozier 1953: 53–71). The larger utilities hired engineering management firms to carry out the conversion tasks. Some utilities, however, managed their own transitions rather than hiring outside contractors, forming schools and training men specifically for the job and relying upon experience from other utilities.[6]

It is informative to compare the transition in the United States to that in Great Britain (Williams 1981: 180–204). The British industry had the advantage of centralization, since it had been nationalized after the Second World War. Conversion was possible because of the discovery of North Sea gas, and the British Gas Council had committed itself to the substitution in 1966, estimating that it would involve over 30 million appliances. In 1965, the Gas Council experimented with a pilot conversion scheme on Canvey Island, located in the Thames about 30 miles from London. The conversion was successful, and the Gas Council, operating through a Conversion Executive, set about replacing manufactured gas with natural gas throughout Great Britain. Subcontractors, operating for a fee or lump sum payment, formed the basis of the labor force for the conversion, with thirteen training schools jointly formed by the Gas Council and the relevant unions. To hold the labor force for the term of the conversion, workers were provided with a terminal bonus at the end of their work. The entire program was estimated to cost nearly £1,000 million when the cost of writing off gas manufacturing plant was included. The British conversion took approximately ten years and involved almost 13 million domestic customers. This compared to conversion times of six years for the Dutch (2 million users) and twelve years for the Japanese (5 million users). A high British governmental official described the conversion as 'perhaps the greatest peacetime operation in this nation's history'.

It is difficult to provide a direct comparison between the British conversion and that in the United States. The US conversion took place over approximately thirty to forty years, with the change-over of the largest utilities occurring in the 1940s and 1950s. The US conversion was carried out by many separate utilities rather than by one central authority operating through regional boards. Another sharp difference involved the pace of the change. A substantial number of US utilities converted to mixed gas rather than directly to natural gas, stretching the conversion process out, whereas the British shifted entirely to natural gas in a single operation. The British conversion took place after most gas utilities in the United States had already switched to natural gas, and the British benefited from the US experience.[7] It is difficult not to conclude that, because of centralization and assured long-term reserves of natural gas, the British conversion was carried out at a lower cost and more expeditiously than that in the US. On the other hand, the use of mixed gas enabled some US utilities to get maximum use out of their manufactured gas technology.

Conclusions

This chapter has focused on the development of the gas industry in the United States. Like other large technical systems, the manufactured and natural gas industries experienced, to use the terminology of Thomas P. Hughes (1983), a number of 'reverse salients'. Those pertaining to transportation, production and use, as well as a lag in regulatory definitions, have been discussed in this essay. As with electricity, load factors and competition from other energy forms played critical roles, driving the industry, sometimes reluctantly, towards innovations such as carbureted water gas and the Welsbach mantle that removed some of the reverse salients.

In order to meet competition, gas utilities also slowly shifted into new product lines such as stoves and water heating, and they utilized innovative methods of merchandising and advertising to sell the appliances in an increasingly consumption-oriented society. The manufactured gas industry also had the prospect of profiting from the sale of byproducts or externalities, an option not possessed by electricity. Thus, markets for byproducts could determine technological shifts and prospects of profit or loss, as well as causing integration with other industries such as the byproduct coke and iron and steel industries.

Manufactured gas systems shared a number of characteristics with other urban networks such as water supply, sewerage, and district heating. As Arne Kaijser has noted, these were 'grid-based systems' that required a special physical network devoted to supplying users (Kaijser 1993: 195–6). They were also centralized rather than decentralized, and were close to being natural monopolies, a characteristic which affected their rate making and consumer relations and, at times, deeply involved them in politics. The substitution of natural for manufactured gas resulted in the gradual death of the manufactured gas industry as a productive entity. However, much of the original industry's grid remained in the form of inner-city distribution systems (the skeletal remains).

From the beginning manufactured gas firms required municipal franchises for their operation because they utilized city streets for their distribution systems. As the industry grew and changed, its activities were increasingly subject to regulation not only by municipal commissions and regulations but also by state commissions. State legislatures originally established these commissions to regulate railroad rates, but they soon extended to other utilities. In 1885, the Commonwealth of Massachusetts established the Massachusetts State Board of Gas Commissioners. By 1906, twenty-nine states had regulatory commissions, with various types of powers. After 1907, however, a new administrative regime began. New York and Wisconsin created the first commissions with the powers to effectively regulate utilities in that year, and they were followed by nearly two-thirds of the states in the next six years. In twenty-eight states these bodies were appointed by the governor, and in fourteen they were elected.

While some of these commissions attempted to tightly control gas utility activities, to a large extent state regulators favored producers' interests more than did municipal regulators. As Werner Troesken notes, 'gas companies preferred no regulation over state regulation, and state regulation over municipal regulation, while consumers and local politicians preferred municipal regulation over state regulation, and state regulation over no regulation' (Troesken 1994: 108–9). The new state commissions promised to provide 'scientific and expert' methods of regulation, insuring consumers fairness in rate setting and removing the public utilities from politics, although this goal was not always clearly met (Anderson 1981: 34–56). With the further expansion of gas grids across state boundaries, pipeline companies became subject, after the Natural Gas Act of 1938, to federal control, which had a constraining and shaping effect on the industry.

The conversion of the US manufactured gas system to a natural gas system was a formidable undertaking. In her introduction to *Changing Large Technical Systems*, Jane Summerton (1994) observes that historically large sociotechnical systems have undergone three types of 'reconfiguration' (Summerton 1994: 5–6): the first is the territorial expansion and interconnection of similar systems over political borders; the second is linkages between systems with heterogeneous functions; and the third involves shifting systems from a basis of monopoly to one of competition and open access. The case of manufactured and natural gas, however, presents still a fourth type of reconfiguration – one in which the delivery system and appliances related to the first system are retained for use by a different but still related form of energy. This situation bears some similarity to the manner in which Samuel Insull in Chicago coupled the old DC electrical technology to the new AC in a 'universal supply system' (Hughes 1983: 208–9).

Finally, the conversion process itself, which meant adapting the distribution system and home appliances to the characteristics of natural gas, as well as accustoming consumers to the new fuel, represented a major organizational challenge. Characteristically, given its decentralized economy and corporate system, the American industry made this shift gradually and in several different stages, learning from experiences over time. Here again, regulatory rulings as well as perceptions of supply adequacy and prospective demand, often shaped the process.

Acknowledgements

I am indebted to the attendees at the Conference in Large Technical Systems held at Autun in September 1995 for their helpful comments. I have especially benefited from the comments of Arne Kaijser. I would also like to thank Christopher James Castaneda for sharing information about the literature of the gas industry.

Notes

1 Here I am using Bernward Joerges' definition of technical systems as 'systems of machineries [sic] and freestanding structures performing, more or less reliably and predictably, complex standardized operations by virtue of being integrated with other social processes, governed and legitimated by formal, knowledge-intensive, impersonal rationalities' (Joerges 1988: 23–4).

2 Manufactured gas plants often left coal tar residues on site, requiring cleanups (see Harkins *et al.* 1984).

3 Ammonia was regarded as a nuisance until after the Civil War when ammonium sulphate was used for fertilizer. Gasworks themselves burned byproduct coke

under their retorts and sold some on the open market but, because of combustion problems, coke had limited appeal as a fuel except in Pittsburgh where it was used in iron-making.

4 A similar development took place in the Netherlands: see Kaijser (this volume). In contrast, few British gas companies operated coke ovens (Williams 1981: 59–60).

5 Castaneda (1993: 120–43) has an excellent chapter dealing with the conversion process in New York city; see also Tussing and Barlow 1984.

6 Companies that converted in the 1920s and early 1930s largely appeared to have handled conversions on their own without advice from management firms (see, for instance, Buckley 1932: 6–24).

7 C.V. Kroeger of Esso (Standard Oil of New Jersey), who had worked with American and European conversions, advised the British Conversion Manager, W.D. Ellis, at the beginning of the process (Williams 1981: 181).

References

Abell, H.C. (1929) 'Changing over a city's gas supply', *Natural Gas* 10: 36–71.

American Gas Association (1956) *Historical Statistics of the Gas Industry*, New York: AGA.

Anderson, Arthur H. (1936) 'Natural gas comes to Detroit: Appliance conversion', *Gas Age Record and Natural Gas* 78: 703–4.

Anderson, Douglas D. (1981) *Regulatory Politics and Electric Utilities: A Case Study in Political Economy*, Boston (MA): Auburn House.

Anderson Jr., Nils, and DeLawyer, Mark W. (1995) *Chemicals, Metals and Men: Gas, Chemicals and Coke: A Bird's Eye View of the Materials That Make the World Go Around*, New York: Vantage Press.

Begeebing, W.C. (1931) 'Obviating changeover troubles', *American Gas Journal* 135: 60–1.

Benjamin, Park (ed.) (1895) *Appleton's Cyclopaedia of Applied Mechanics*, New York: D. Appleton and Co.

Binder, Frederick Moore (1974) *Coal Age Empire: Pennsylvania Coal and its Utilization to 1860*, Harrisburg (PA): Pennsylvania Historical and Museum Commission.

Blake, A.E. (1922) 'Water gas', in Raymond Foss Bacon and William Allen Hamor (eds) *American Fuels* (Vol. 2), New York: McGraw-Hill.

Block, Eugene B. (1966) *Above the Civil War: The Story of Thaddeus Lowe, Balloonist, Inventor, Railway Builder*, Berkeley (CA): Howell-Nort Books.

Brown, James A. (1931) 'What it means to change over to natural gas', *Gas Age Record* 28: 161–6.

Buckley, William J. (1932) 'Chicago's change-over from manufactured to mixed gases', *Natural Gas* 13: 6–24.

Business Week (1951) 'Con Edison turns on the gas', 23 June: 108–13.

Castaneda, Christopher James (1993) *Regulated Enterprise. Natural Gas Pipelines and Northeastern Markets 1938–1954*, Columbus (OH): Ohio State University Press.

Castaneda, Christopher James and Pratt, Joseph A. (1993) *From Texas to the East: A Strategic History of Texas Eastern Corporation*, College Station (TX): Texas A&M University Press.

Clark, Walton (1916) 'Historical notes on a century of development in lighting', *Gas Age* 38: 665–8.

Cochrane, Rexmond C. (1966) *Measures for Progress: A History of the National Bureau of Standards*, Washington (DC): US Dept of Commerce.

Collins, Frederick L. (1934) *Consolidated Gas Company of New York*, New York: the company.

Commonwealth of Pennsylvania (1875) *Second Geological Survey of Pennsylvania: 1875. Special Report on the Coke Manufacture of the Youghiogheny River Valley in Fayette and Westmoreland Counties*, Harrisburg, (PA): Board of Commissioners for the Second Geological Survey.

Davis, Ralph E. (1935) 'Natural gas pipe line development during the past ten years', *Natural Gas Magazine* 6: 3–4.

Du Puy, William Atherton (1930) 'Natural gas, our new bonanza', *World's Work* 59: 55–8, 98.

Eavenson, Howard N. (1942) *The First Century and a Quarter of American Coal Industry*, Pittsburgh (PA): privately printed.

Elton, Arthur (1958) 'Gas for light and heat', in Charles Singer *et al.* (eds) *A History of Technology, Vol. IV: The Industrial Revolution c. 1750–1850*, New York: Oxford University Press.

Ershkowitz, Henry B. (n.d.) 'History of the Philadelphia gas works: a century and a half of progress 1836–1986', unpublished manuscript, Temple University, Philadelphia (PA).

Gas Age (1919) 'Coke plant producing gas for domestic purposes', 43: 11–12.

—— (1921) 'To abolish candle power standard', 47: 216.

—— (1937) 'Grand Rapids permitted to serve unmixed gas', 80–6.

—— (1947) 'Natural gas comes to Washington', 99: 26–8.

Gas Engineers Handbook: Fuel Gas Engineering Practices (1977), New York: Industrial Press.

Gilmer, Robert W. (1981) 'The history of natural gas pipelines in the southwest', *Texas Business Review* 1: 129–32.

Grozier, David (1953) 'The Brooklyn Union natural gas conversion: biggest changeover in the world', *Gas Age* 111: 39–41.

Harkins, Scott M. *et al.* (1984) 'US production of manufactured gases: assessment of past disposal practices', unpublished EPA Report, Research Triangle Park (NC): Research Triangle Institute.

Hasiam, Robert T. and Russell, Robert P. (1926) *Fuels and Their Combustion*, New York: McGraw-Hill.

Hughes, Thomas P. (1983) *Networks of Power: Electrification in Western Society 1880–1930*, Baltimore (MD): Johns Hopkins University Press.

Hunt, Arthur L. (1900) 'Gas, Manufactured', in *US Bureau of the Census, Twelfth Census of Manufacturers (1900) Part IV: Reports on Selected Industries*, Washington (DC): GPO.

Jacobson, Charles D. and Tarr, Joel A. (1995) 'Ownership and Financing of Infrastructure: Historical Perspectives', *Policy Research Working Paper* 1466, Washington: The World Bank.

Joerges, Bernward (1988) 'Large Technical Systems: Concepts and Issues', in Renate Mayntz and Thomas P. Hughes, (eds) *The Development of Large Technical Systems*, Boulder (CO): Westview Press.

Johnson, George B. (1949) 'Gas conversion experiences', *Gas Age* 104: 38–43, 62–70.

Johnson, Leland (n.d.) *The Headwaters District: A History of the Pittsburgh District*, US Army Corps of Engineers, Washington (DC): GPO.

Kaijser, Arne (1990) 'City lights: the establishment of the first Swedish gasworks', *Flux, Cahiers scientifiques internationaux Réseaux et Territoires* 1: 77–84.

—— (1993) 'Fighting for lighting and cooking: competing energy systems in Sweden 1880–1960', in William Aspray (ed.) *Technological Competitiveness: Contemporary and Historical Perspectives on the Electrical, Electronics, and Computer Industries*, New York: IEEE Press.

King, Thomson (1950) *Consolidated of Baltimore 1816–1950*, Baltimore (MD): the company.

Leinroth, J.P. (1928) 'Industrial gas in the United States: Growth and future trends', in *Transactions of the Fuel Conference, Vol. II: The Carbonisation Industry Utilization of Fuels*, London: Percy Lund, Humphries & Co.: 1211–17.

Littlefield, Douglas R. and Thorne, Tanis C. (1990) *The Spirit of Enterprise: The History of Pacific Enterprises from 1886 to 1989*, Los Angeles (CA): Pacific Enterprises.

Matthews, Derek (1987) 'The technical transformation of the late nineteenth-century gas industry', *Journal of Economic History* 47: 970–80.

McGannon, Harold E. (ed.) (1964) *The Making, Shaping and Treating of Steel*, Pittsburgh (PA): US Steel, 8th edn.

Moehring, Eugene P. (1981) *Public Works and the Patterns of Urban Real Estate Growth in Manhattan 1835–1894*, New York: Arno Press.

Parr, S.W. (1932) *The Analysis of Fuel, Gas, Water and Lubricants*, New York: McGraw-Hill.

Passer, Harold C. (1953) *The Electrical Manufacturers 1875–1900*, Cambridge (MA): Harvard University Press.

—— (1967) 'The electric light and the gas light: innovation and continuity in economic history', in Hugh G.J. Aitken (ed.) *Explorations in Enterprise*, Cambridge (MA): Harvard University Press. (I am indebted to David Hounshell for bringing this essay to my attention.)

Plata, Richard A. and Ferrarini, Joseph (1949) 'Advance organization: detailed planning speed Staten Island conversion job', *American Gas Journal* 171 (Sept.): 8–11.

Platt, Harold L. (1991) *The Electric City: Energy and the Growth of the Chicago Area 1880–1930*, Chicago: University of Chicago Press.

Rhodes, Edmund O. (1966) 'The history of coal tar and light oil', in Arnold J. Hoiberg (ed.) *Bituminous Materials: Asphalts, Tars, and Pitches, Vol. III: Coal Tars and Pitches*, New York: Interscience Publishers.

Rose, Mark H. (1995) *Cities of Light and Heat: Domesticating Gas and Electricity in Urban America*, State College (PA): Pennsylvania State University Press.

Rumer, Thomas A. (1983) *Citizens Gas & Coke Utility: A History 1851–1980*, Indianapolis (IN): the company.

Sanders, M. Elizabeth (1981) *The Regulation of Natural Gas Policy and Politics 1938–1978*, Philadelphia (PA): Temple University Press.

Schivelbusch, Wolfgang (1988) *Disenchanted Night: The Industrialization of Light in the Nineteenth Century*, Berkeley (CA): University of California Press.

Singer, Charles and Williams, Trevor I. (1958) *A History of Technology Vol. IV The Industrial Revolution c. 1750–1850*, New York: Oxford University Press.

Steere, F.W. (1922) 'Producer gas technology', in Raymond Foss Bacon and William Allen Hamor (eds) *American Fuels (Vol. 2)*, New York: McGraw-Hill.

Stotz, Louis and Jamieson, Alexander (1938) *History of the Gas Industry*, New York: Stettiner Bros.

Summerton, Jane (ed.) (1994) *Changing Large Technical Systems*, Boulder (CO): Westview Press.

Tarr, Joel A. and Dupuy, Gabriel (1988) *Technology and the Rise of the Networked City in Europe and America*, Philadelphia (PA): Temple University Press.

Thoenen, Eugene D. (1964) *History of the Oil and Gas Industry in West Virginia*, Charleston (WV): Education Foundation Inc.

Troesken, Werner (1994) 'Why regulate utilities: an historical perspective', unpublished manuscript, Pittsburgh (PA).

Tryon, F.G. (1928) 'Domestic fuels other than anthracite. Part I: processed coals', in *Transactions of the Fuel Conference, Vol. II: The Carbonisation Industry Utilization of Fuels*, London: Percy Lund, Humphries & Co.: 333–4.

Tussing, Arlon R. and Barlow, Connie C. (1984) *The Natural Gas Industry: Evolution, Structure, and Economics*, Cambridge (MA): Ballinger Pub. Co.

US Bureau of the Census (1870) *Ninth Census (1870), Vol. III: Manufactures*, Washington (DC): GPO.

—— (1889) *Eleventh Census of Manufactures (1890), Part III: Selected Industries, Gas*, Washington (DC): GPO.

—— (1895) 'Gas', in *Eleventh Census of Manufactures, Part III: Selected Industries*, Washington (DC): GPO.

—— (1900) *Twelfth Census of Manufactures, Part IV: Selected Industries*, Washington (DC): GPO.

—— (1921) *Fourteenth Census of the US (1920), Vol. X: Manufacturers, Report for Selected Industries*, Washington (DC): GPO.

—— (1930) 'Gas, Manufactured, Illuminating and Heating', *Fifteenth Census (1930), Manufacturers, Vol. VII: Reports on Industries*, Washington (DC): GPO.

Wainwright, N.B. (1961) *History of the Philadelphia Electric Company 1881–1961*, Philadelphia (PA): Philadelphia Electric Company.

Weber, F.C. (1922) 'The future of the artificial gas industry' in Raymond Foss Bacon and William Allen Hamor (eds) *American Fuels*, New York: McGraw-Hill.

Wehrle, George (1931) 'Effect of change-over to natural gas', *Oil and Gas Journal* 29: 155–6.

Williams, Trevor I. (1981) *A History of the British Gas Industry*, Oxford: Oxford University Press.

Williamson, Harold F. and Daum, Arnold R. (1959) *The American Petroleum Industry: The Age of Illumination 1859–1899*, Evanston (IL): Northwestern University Press.

3

STRIKING BONANZA

The establishment of a natural gas regime in the Netherlands

Arne Kaijser

Introduction

On 22 July 1959 an exploration team drilling near the village of Slochteren found a natural gas reservoir. After further drilling in the following year, it turned out to be a very big reservoir, in fact the biggest reservoir in western Europe at the time. Moreover it was ideally situated: the major urban centres of West Germany, France and Great Britain, as well as the whole of Belgium and the Netherlands were situated within a radius of 500 miles. In short, they had struck a bonanza.[1]

The question of how to organize the exploitation of this gas field in the province of Groningen became a central political issue in the Netherlands in the early 1960s. It was soon clear that the existing organizational and legal framework within the Dutch gas industry was ill-adapted to this task. There was need for a new regime in the industry. I use the word 'regime' (Thue 1995) to refer to the legal framework, the organizational structure and the rules of the game within a given industry in a given country. This new regime, specifying new roles and responsibilities for the major organizations in the gas sector, was created in just two and a half years. Furthermore, an elaborate strategy was developed to build a national grid interconnecting the Groningen field with existing regional and local gas systems, and to market the gas. The new regime and strategy paved the way for the rapid arrival of Groningen gas on the European market: ten years later almost half of the Dutch energy supply was based on this new source. In addition a large share of the gas was being exported to neighbouring countries.

This chapter focuses on the critical years in the early 1960s when the new regime was elaborated and implemented. It analyses the problems encountered, the objectives and visions of the various actors involved, and the processes through which consensus was reached. It then discusses how the new regime contributed to the development of a national natural gas system and how it coped with the oil crisis.

The chapter thus deals primarily with institutional issues, rather than technological ones. The reason is that I regard the then existing gas regime as the major 'reverse salient' (Hughes 1983) hindering the large scale introduction of Groningen gas. The technical issues were not as critical, because technology for piping gas over long distances had been developed decades earlier in the US (see Joel Tarr's article in this volume), even though it had to be adapted to Dutch conditions. More generally, I believe that institutional issues frequently were of critical importance for the geographic interconnection of infrastructural systems, not the least for electricity and the telephone, which just like gas have grown from local networks to international systems (Kaijser 1987, 1992, 1995).

The Groningen field and the need for a new gas regime

Historical background

Gas technology was one of the epochal innovations of the nineteenth century, making it possible to transform cheap fuels like coal, peat and wood to a flexible and high-quality energy carrier. This new technology was developed in Britain in the early nineteenth century for the lighting of large factories. The first public gasworks was built in London in 1813. The transfer of gas technology to the European continent and the United States started in the 1820s (Falkus 1967; Tarr, this volume).

Amsterdam and Rotterdam were among the earliest continental cities to obtain gasworks. These gasworks, built in the mid-1820s, were privately owned and used British technology. Between 1840 and 1870 a rapid diffusion of gas technology occurred in the Netherlands and by 1870 there were more than a hundred gasworks throughout the country. The early gasworks were all privately owned, but from the mid-nineteenth century municipal ownership was introduced and gradually became the dominant mode. By the turn of the century a homogeneous regime had developed in the gas industry. Gas supply was seen as a public service that ought to be carried out by local authorities under monopoly conditions. I call this the *municipal regime* (Dil and Homburg 1993).[2]

In the 1930s a new phase in the development of gas systems started in the Netherlands implying a gradual transition from local gasworks to regional systems. The rationale for building regional systems was to substitute locally produced gas with cheaper 'distance-gas' as it was called. Based on gas from coke-factories, the first regional systems were developed from the 1930s by DSM (the Dutch State Mine, a state owned coal company) and Hoogoven (a partially state owned steel company). The two systems gradually expanded, and by the late 1950s the DSM grid covered most of the country south of the Rhine and the Hoogoven grid covered the whole province of North Holland (Nota 1953).

In the early 1950s two new kinds of 'distance-gas' were introduced that were not based on coal. One was gas made from petroleum. It was produced as a by-product at a new oil refinery in Rotterdam harbour, built by Shell in 1950. The other was natural gas, extracted at a number of small wells primarily in the north-eastern part of the country. This gas was produced by the Dutch Oil Company, NAM, a company founded in 1947 and owned jointly by the Royal Dutch Shell Group ('Shell'), and Standard Oil of New Jersey ('Esso'). NAM had obtained a concession from the Dutch State to produce oil in the province of Drenthe.

In 1948 NAM discovered natural gas within its concession area. In accordance with the traditional rules in the industry, the transportation and sales of gas were regarded as a public utility. Thus, the state added a special clause to NAM's concession contract, specifying that NAM had to sell all natural gas to the state for a 'reasonable price'. A state gas company, SGB, was established in 1954, for the purpose of buying natural gas from NAM, piping it and selling it to municipal gasworks. In the mid-1950s SGB therefore built regional gas grids, starting in the northeastern part of the country, then also in the area around The Hague, where some gas deposits were found too (Kielich 1988: 16f, Nota 1962: 2).

Both natural gas and refinery gas had different chemical compositions from traditional town gas, both with a much higher calorific value. This implied that they did not fit the existing gas networks. There were two possibilities to overcome this mismatch: to adapt the gas to the networks; or to adapt the networks to the gas. The first was achieved by lowering the calorific value of the gas, for example through blending it with a low calorific gas such as nitrogen. The other possibility was to convert all appliances to the new gas, by changing the proportion of gas and air in every burner.[3] The former solution prevailed (Nota 1958).

The development of regional gas systems from the 1930s implied profound institutional changes. The municipal regime in the gas industry was replaced with a new, *regional regime*. Local distribution was still carried out by municipal gasworks, but production and transmission were for the most part transferred to new players. These included very different kinds of organizations – a coal mining company, a steel company, oil companies and a state transportation authority – which had different motives for their gas business. The new regional regime was thus characterized by organizational diversity. Technically too, gas systems were more heterogeneous than before, with several kinds of gas being used. This double heterogeneity was an important reason for the failure in the 1950s of several attempts to achieve the national integration of regional gas grids in the Netherlands (Nota 1958).

The silence of Slochteren

Let us now return to where we started – the village of Slochteren in July 1959 and the discovery of what became known as the Groningen gas field. When NAM realized what they had discovered they did not triumphantly announce it to the world, but tried to keep it secret as long as possible. The reason seems obvious: NAM had a concession for exploration but not for exploitation in the province of Groningen, and they wanted to explore the size of the new field and possibly neighbouring ones in order to apply for a production concession. And they did not want competing oil companies entering the scene.

The 'silence of Slochteren' lasted for more than a year. Very few people seem to have been informed during that period, and those who knew kept quiet. On 14 October 1960 the silence was finally broken. In a debate on energy policy in the European Parliament in Strasbourg, a Belgian politician, V. Leeman, revealed that a huge natural gas field containing 300 billion cubic metres, equivalent to 300 million tons of coal, had been found in the north of the Netherlands. This sensational announcement appeared in all major business journals (Kielich 1988: 23f, Salverda and Bakker 1983: 7).

Four days after Leeman's announcement, NAM wrote a letter to the Minister for Economic Affairs, J.W. De Pous, to inform him that the gas field in Slochteren seemed bigger than had previously been thought, with an estimated size of 60 billion cubic metres. NAM also requested a concession over the new gas field: they wanted to produce, transport and sell a major share of the gas directly to large customers such as industries and power companies without SGB as a go-between (Kielich 1988: 31).

This letter was the starting point of an intensive political negotiation process which lasted for two and a half years and resulted, in the Spring of 1963, in the creation of a new regime for the Dutch gas sector. Discussions frequently involved only a few people and took place behind closed doors at the Ministry for Economic Affairs or at the Shell headquarters. On a few occasions, however, gas policy issues were discussed openly in Parliament.

The form of these negotiations was partly a reflection of a fundamental characteristic of the Dutch political system in the early 1960s, namely its segmentation into various religious or political blocs, which are known in Dutch as 'zuilen' ('pillars' or 'columns'). The word suggests that Dutch society was traditionally built upon a number of 'pillars': one Protestant, one Catholic, one socialist and one liberal/conservative.[4] The key feature of 'pillarization' was that each pillar nearly entirely organized the social life of its members. Ordinary people had little contact with members of other pillars. But to avoid conflicts, the pillars' respective elites developed a spirit of consensus-seeking. Because each pillar represented a minority, none could make decisions without support from the others. In order to reach consensus, major decisions were often prepared behind closed doors.

In what follows, I will first present the key players and the critical issues, and then I will briefly sketch some of the major events in the negotiation process.

The players

The principal player in the negotiation process was the Minister for Economic Affairs, J.W. De Pous. He had been appointed in 1959 and was young and somewhat inexperienced for a top politician. De Pous belonged to the Protestant pillar, but in economic matters he could be described as a liberal. He belonged to De Quay's cabinet (1959–63), a broad cabinet including the Protestant, Catholic and Liberal pillars (Salverda and Bakker 1983: 13f).

As the Minister for Economic Affairs, De Pous had considerable power. His general responsibilities covered industrial and energy policy. He had a strong influence over several important players in the energy sector. The State Gas Company, SGB, was placed directly under his control. He also had strategic influence over the Dutch State Mines, DSM, even though this company had more autonomy than SGB. And his trump card in the negotiation process was that, in accordance with Dutch mining law, he was responsible for granting the production concession for natural gas that NAM wanted to secure.[5] De Pous realized at an early stage that the Groningen gas matter was the biggest issue that his Ministry had ever handled. He chose to involve only a handful of staff members at the Ministry to minimize the risk of important information leaking out. His closest collaborator in the gas matter was L.G. Wansink, who was responsible for daily contact with the other players throughout the negotiations.[6]

L. Schepers was De Pous's major counterpart on the industry side. Schepers was one of the managing directors of Shell, in charge of gas. This made him the main representative of the NAM partners. As one of the richest companies in the world and more than half-owned by Dutch capital, Shell played a prominent role in Dutch economic and political life. As a consequence, Schepers was one of the most influential industrialists in the country.

NAM was owned jointly by Shell and Esso. The history of this company actually goes back to 1932, when Shell and Esso made an agreement to start exploration for oil on a joint venture basis in the Netherlands and in Cuba. The main purpose of this agreement was to spread financial risks equally. Esso explored Cuba (where no field was found) and Shell explored the Netherlands. In 1933 it acquired a concession to carry out explorations in different parts of the Netherlands without much success at first. Shortly after the war, oil was discovered in several places in quantities large enough for commercial purposes. Esso then reminded Shell of the 1932 agreement, and in 1947 NAM was established on an equal share basis for future exploration and exploitation of oil:[7] Shell would be the operating company, while Esso would provide the capital and drilling equipment – both scarce resources in postwar Europe (Kielich 1988, Borghuis 1988).

In the 1950s NAM was quite a small business. But with the discovery of the Groningen field, NAM became of strategic importance for the two oil giants. With the pooled resources of Shell and Esso behind it, NAM held a strong position: both Shell and Esso had enormous resources in terms of capital and manpower, and they had a lot of experience of oil politics on a global scale (Yergin 1991), even though their knowledge of natural gas was more limited. But even if NAM and its two parent companies had strong positions in many respects, they lacked two critical resources: a production concession for the Groningen field and a grid to transport the gas to potential customers.

There were also other organizations angling for prominent roles in the emerging gas industry. One of them was the state gas company SGB. In comparison with the other companies involved, SGB was very small, with only about 200 employees. But it controlled an important resource: a gas grid covering almost half of the country. It also had a strategic position in the existing gas market, as the sole authorized buyer of all natural gas produced by NAM.

Another active organization was the Dutch State Mines DSM. As we have seen above, DSM was the gas supplier in the southern part of the country. But DSM was primarily the major coal producer in the country, and with 45,000 employees it was by far the largest company in the province of Limburg. Even though DSM was a state owned company, its management had a strong position *vis-à-vis* the ministry because the company was big and because it had strong backing from provincial authorities (Messing 1988).

One more player has to be taken into account, although it did not participate a great deal in the direct negotiations until the very end: the Social Democratic Party (PvdA). The PvdA was the major political representative of the socialist pillar in Dutch society. Although it was not represented in the cabinet in the early 1960s, a stable long-term solution to the gas matter would have to be supported by the PvdA. The traditional position of the PvdA was that public utilities should be entirely controlled by public authorities at the state, provincial or municipal level. This position was in almost total contradiction with the demands put forward by NAM in its letter to De Pous.

Finally, it is also interesting to note that there were organizations within the traditional gas industry that did *not* take an active part in the process. Particularly, the municipal gasworks were largely passive in the early 1960s. Naturally, they hoped that the Groningen gas field would provide them with cheap gas, but they did not make a coordinated effort to influence the outcome of the negotiations. Only in 1963, when the new regime was already in place, did they create a joint organization, SROG, to protect their common interests.

Marketing the Groningen gas and constructing the new gas regime

Two key issues had to be addressed in the negotiation process. The first concerned the future regime of the gas industry. It had two main aspects: Which existing or new organizations should participate in the sector and what would their roles and responsibilities be? How should the legal and economic framework, the 'rules of the game', be designed? The second key issue concerned the strategy for selling the gas. How should the gas be commercialized to provide the highest benefits to all parties? These two issues were discussed in tandem throughout the process, but the emphasis changed over time. In the first few months after the silence of Slochteren had been broken, attention was focused on the strategy issue.

A radical marketing plan

Esso had had a passive role in the NAM operations in the 1950s. Almost no one at the headquarters in New York knew about the company. According to a story told by a former Esso employee, the president of Esso read about Leeman's statement concerning the Groningen field in the *New York Times*, and immediately shouted to his assistant: 'I'll buy it all!' Half an hour later his assistant informed him: 'We already own half of it'. The president took immediate measures, and within forty eight hours two of the company's leading gas experts had arrived in the Netherlands. The two experts had just made a very critical evaluation of Esso's strategy in the US gas industry, which consisted of producing gas and selling it at a low price to 'pipeline companies'. They intended to foster a totally different strategy in the Netherlands.[8]

Working together with two Dutch Esso employees, they identified three major potential markets for natural gas, at very different price levels: the existing gas market, primarily for domestic cooking, at a price level of over 30 cents per cubic metre; space heating, then primarily supplied with coal and oil, at prices between 10 and 15 cents per cubic metre; industrial furnaces and electrical power plants, at prices between 4 and 7 cents per cubic metre.[9] The conventional wisdom then prevailing in the gas industry was that the best way to market gas from large new fields was to sell it all to large industrial customers. The Esso team elaborated a totally new marketing plan with two objectives: achieving a rapid introduction of the Groningen gas, and maximizing the profits from gas sales.

The corner stone of Esso's marketing plan was to use most of the gas for space heating: a household converting to gas heating would increase its gas consumption nearly tenfold. None the less, the Esso team believed that it would be possible to use the existing distribution networks for this increased supply: the Groningen gas had twice the energy content of town gas and pressure could be increased. Furthermore, the load curve for heating was more even than for cooking. Using existing networks for space heating would mean small additional distribution costs. However, a prerequisite was of course that a new high-pressure gas grid be built to transport large quantities of gas to all parts of the country, and that a massive conversion of all local distribution networks from town gas to natural gas be carried out, by readjusting all burners and appliances.[10]

Another important element of the marketing plan of the Esso team was that gas should have a new, regressive household tariff. The first 600 cubic metres (sufficient for cooking) would be charged at about 25 cents per unit, but additional quantities would be charged at only about 10 cents per unit. Gas would thus be competitive both for cooking (vs electricity) and for space heating (vs coal and oil). The Esso team argued that the gas industry as a whole would double its income if it adopted this sales strategy rather than the traditional strategy.

This radical marketing plan was presented in a report in December 1960. It was received with great scepticism both by Shell and local gas companies. The critics argued that it was unrealistic to expect a large scale conversion to gas heating, not least because most households could not afford to make a big investment in a new heating appliance. And, they added, if such a conversion took place none the less, it would be impossible to supply enough gas for heating on cold winter days. Shell presented an alternative plan a month later, according to which 90 per cent of the gas would be sold to industrial customers.

The Esso team did not give up, however. They made a very detailed study of potential conversions to gas heating in Hilversum, a middle sized Dutch town, and then wrote a new report. This time they managed to convince most of their opponents. In March 1961, Shell and Esso presented the plan to De Pous (Kielich 1988: 33f). De Pous and his colleagues had discussed the issue of how to market the gas for some months, and they had reached conclusions similar to those of the Esso team.[11] De Pous thus agreed with the basic principle in the Esso report: to maximize income from gas sales by primarily selling gas in those sections of the market where prices were higher. This pricing policy was believed to bring about the maximum benefit for the national economy as a whole. The same approach (profit maximization) was used for gas exports when it became clear that the domestic market would not 'swallow up' all of the Groningen gas. This may seem an obvious commercial policy, but there was an alternative, namely to supply households (voters) and domestic industries with cheap gas, as was done in Britain in the late 1960s (Davis 1984: 160 and ch. 5).

There seems to have been another reason for De Pous's support for this pricing policy, a reason that had to do with energy policy at large. A price-dumping policy would probably have led to big disturbances in the existing energy market and a fast aggravation of the difficulties of the coal mining industry. De Pous was aware of that and he insisted that the

Groningen gas should be introduced as smoothly as possible. The pricing policy can be seen as one key element in this process (Nota 1962: 4, 10, Davis 1984: 157).

Furthermore, De Pous agreed with the objective of rapidly expanding gas sales. There were two main arguments for speed. The first was the traditional reason in capital-intensive enterprises: enormous investments would have to be made to build production and transportation facilities, and money was needed quickly to pay off loans. The second was more specific to the early 1960s. The Netherlands, like many other countries, supported large R&D programmes in nuclear energy at the time, and there was a common belief that in due time nuclear energy would become a low price competitor to gas. This made it essential to exploit as much of the Groningen gas as possible before cheap nuclear energy became available. Obviously, this expectation of falling energy prices was also an incentive for all parties to reach a solution to the gas question as soon as possible.[12]

The organizational set-up

While Esso and Shell were focusing on marketing options, De Pous and his staff examined institutional matters. De Pous soon realized that the contract between SGB and NAM, an important element of the existing regime, would have to be changed. The contract stipulated that NAM had to sell all its natural gas to SGB at a 'reasonable' price (only 2 cents per cubic metre for volumes above half a million cubic metres per day), and that SGB had to pay for all the gas that NAM offered to sell each year, even if it could not resell all of it. In the early 1960s, SGB sold only half the gas that NAM offered and had to pay a sizeable amount of money for gas it couldn't make use of. De Pous and his staff realized that this contract was in fact a threat to the state. If NAM included the Groningen gas in their yearly offer to SGB, SGB would be unable to pay for it. The oil companies were aware of this potential weapon. The Shell management apparently joked about declaring the state bankrupt if SGB didn't pay (Salverda and Bakker 1983: 13).

De Pous and his staff came to the conclusion that the existing arrangements were unsatisfactory because there was no coordination whatsoever between NAM and SGB. They thus agreed with NAM that there was a need for an integrated organizational structure, in which production, transportation and sales of gas would be closely coordinated. Such a coordination was of particular importance given the huge investments that would have to be made. But the ministry disagreed with NAM on who would own and control these new organizations.

The original request by NAM (in their letter dated 18 October 1960) to produce, transport and sell most of the gas, was totally unacceptable to De Pous, as it would mean abandoning the old principle of public influence in the gas industry. He was, however, prepared to accept NAM's participation for several reasons. One was an attitude of 'fair play': it was NAM that had discovered the new gas field after many years of exploration. Another reason was pragmatism: NAM's financial and human resources were of vital importance if a large-scale gas system was to be developed quickly. A third reason was that with the exploitation of the Groningen gas field, gas would partly lose some of its public utility character. Just like electricity, gas had previously been mostly manufactured from coal and considered as a convenient *energy carrier* for a variety of purposes. But with the Groningen field, natural gas became an important *energy source* in its own right, comparable to coal and oil. The available quantities were so large that gas would also have to be used instead of coal and oil for low quality purposes.[13]

De Pous, from an early stage in the negotiation process strove for an organizational solution where the state and NAM would be jointly responsible for the production, piping and sale of gas. Defining precisely the respective prerogatives of NAM and the state was a key issue, to which we shall return below. Another important issue concerned the design of state participation itself. It could not be the ministry alone. The question was then which other state department(s) should be involved? There were two main candidates: SGB and DSM.

SGB was in some ways the natural choice: the company was already responsible for all transportation and sale of natural gas in the country. SGB management repeatedly argued for a strong position for SGB in the future gas industry.[14] However, internal memoranda show that, in March 1961, the minister and his staff came to the decision that SGB would *not* be the state representative. One strong argument against SGB was their small size and limited expertise in comparison with DSM. Or as De Pous later put it in an interview:

> SGB was a company of limited size exclusively operating on the domestic market, not internationally as DSM did. DSM was the best alternative available to the state; they could take on the 'big boys' in the oil industry.
>
> (Bakker and Saverda 1983: 14, my translation)

The decision also included considerations of state energy policy. As mentioned above, the massive input of natural gas in the energy market could have major effects on the future sales of coal and oil. The Limburg coal mines, employing more than 50,000 miners, would be threatened and the rapidly expanding oil industry would be affected. De Pous came to the conclusion that a simple way to avoid destabilization would be to give the responsibility for the future gas industry to the major coal and oil companies in the country, DSM, Shell and ESSO. This consensus-based approach, which led to a merger of the dominant players in the existing energy market has no equivalent in the world of natural gas (Davis 1984: 157).

But all parties were not happy with this set-up. On 5 September 1961, SGB management sent a long letter to De Pous in which they asked 'whether the natural gas interests would be safeguarded if they were entrusted to the oil and coal interests'.[15] However, they had no say: De Pous had made his mind up half a year earlier, in March 1961, and had asked DSM, Shell and Esso to make a joint proposal for the future organization of the gas industry.

Shaping consensus

It took the three companies three months to work out a joint proposal which they presented to De Pous in June 1961. One part concerned the marketing strategy; it followed the lines of the Esso report, which meant that DSM accepted the introduction of gas for space heating on to the market, despite expected adverse effects on coal heating. The other part dealt with organizational structure. The proposal was to create two new companies, one for exploitation and one for transportation and sales on the Dutch market. Shell, Esso and DSM would each own one-third of the shares in each of the two new companies. The coordination between production and sales would be achieved through the personal connections between the boards of the two companies.

De Pous and his staff were not happy with this proposal because it gave only a minority position to the State. They regretted that they had not taken part in the preparation of this proposal.[16] But instead of bluntly rejecting the proposal, De Pous installed an Advisory Committee to look over the proposal and discuss it with all affected parties. The three members of the committee were carefully selected. The chairman, W.C.L. van der Grinten, was Professor of Law, chairman of the Coal Mining Board, former deputy Minister for Economic Affairs and a Catholic Party sympathizer. Board members included H. Vos, vice-chairman of the Coal Mining Board and a Member of Parliament for the PvdA, and T.P. Tromp, former Minister of Transport, a member of the board of Philips and a Liberal Party sympathizer. A member of De Pous's staff, L.G. Wansink, was appointed secretary. Thus Protestant De Pous made sure that the other 'pillars' were represented on the committee; it made the committee an ideal setting for a compromise.

The committee worked fast and in secret. It was appointed in August and presented its confidential report to the minister in December 1961.[17] The committee had auditioned many organizations: for example, the power industry had argued that gas should be used primarily for power generation; elected representatives of the province of Groningen, one of the poorest in the country, asked that at least part of the Groningen gas be sold at a low price to local industries in order to stimulate the economic development of the province (Nota 1962: 5, Salverda and Bakker 1983: 19).

The committee had to address one particularly delicate issue. A year earlier, in September 1960, the five major oil exporting countries Saudi Arabia, Venezuela, Kuwait, Iraq and Iran had established the Organization of Oil

Exporting Countries (OPEC). This was a demonstration of power of these countries *vis-à-vis* the oil companies, and the OPEC founders demanded an increased state participation in the oil and gas production in their countries (Yergin 1991: 519f). Shell and Esso pointed out both to the committee and to De Pous that it would be much more difficult for them to dismiss OPEC's claims if the Netherlands decided for direct state participation in the exploitation of the Groningen field. This issue might also affect gas exports.

Both the minister and the committee showed a positive understanding of this issue and, together with Shell's legal department, the committee worked out a solution: NAM would still hold the concession, but a new financial entity (holding company) called the 'Maatschap' and involving DSM would share the costs and – more importantly – the profits of the business. Another new company called Gasunie would be responsible for transport and sales within the Netherlands. And finally, an organization called NAM/Gas Export would be formally in charge of gas exports. Much of this arrangement, in particular NAM/Gas Export, was a façade hiding the strong influence of the state. In reality Gasunie would be the leading player, and Maatschap and Gasunie would share the same board.[18]

All parties agreed on the general structure, but there was still no consensus about the allocation of shares in Gasunie and in Maatschap. The committee had proposed an equal split between NAM and DSM of the shares in the two new companies, but the oil companies had replied that 60 per cent of the shares for them was as low as they could go. The Committee had brought the negotiation process as far as they could: it was time for De Pous himself to engage directly in the negotiation process.

Parallel negotiations

Following the report of the committee, the ministry started negotiations with Shell and Esso concerning the allocation of power and profits, and with the Parliament concerning the new gas policy as a whole.

De Pous pursued three goals: that the new companies' shares be equally split between NAM and DSM; that his ministry have authority over the conditions of gas deliveries; and that, as a whole, over 70 per cent of the total revenue of gas sales pass to the State (through DSM's share, profit taxation of NAM and concession fees). To attain these goals, De Pous played the PvdA card: a considerable part of the PvdA wanted the complete nationalization of the gas industry, and the party would never accept giving away 60 per cent of the shares to the oil companies. De Pous pointed out that a stable long-term arrangement was dependent on the support of the PvdA, and he assured the oil companies that 'the agreements that I make with you, I will get accepted in Parliament' (quoted by Salverda and Bakker 1983: 32).

Negotiations between the oil companies and the ministry usually involved several high ranking officials from both sides. But, on one occasion, discussions got stuck and De Pous invited Schepers to his office for a tête-à-tête. The minister used the ultimate threat: he told Schepers that if NAM did not agree to his proposal, the State would refuse them the concession for the Groningen field and give them a 'once and for all' compensation package instead. Schepers answered that in that event the refineries and petrochemical plants in the Rotterdam harbour would not be expanded further. Both men now very clearly understood that they were mutually dependent on each other and that an open conflict would be disastrous; they just had to find common ground (see Salverda and Bakker 1983).

In parallel with these negotiations, ministry staff were drafting a natural gas bill, and in March 1962 the draft was presented first to the cabinet, then to the Parliamentary Committee on Economic Affairs. The cabinet agreed, but the Parliamentary Committee's PvdA representatives expressed many criticisms. De Pous did his very best to comply with PvdA's demands, so that they would feel committed to the final bill.[19] On 11 July 1962, the 'Nota inzake het aardgas' was examined in parliament.

The bill described in detail the main features of the new gas regime. As proposed by the van der Grinten committee, NAM would hold the concession and a new financial entity (Maatschap) would be created. Transport and wholesales of gas would be handled by a new company, Gasunie, with a 50/50 state–private ownership. Approximately 70 per cent of the total revenue of gas sales would accrue to the state. In addition the bill proposed that the Minister for Economic Affairs would have important powers regarding nominations to the board of directors of Maatschap and Gasunie, gas tariffs, export and domestic shares in gas sales and discriminatory prices (read: allow extra low prices for industrial customers in the poor northern part of the country, to be limited to 25 million cubic metres per year).

On 4 October, the bill was discussed in the Second Chamber. Governmental parties all praised the minister for an excellent bill and had only minor comments (many wanted to increase the quantity of cheap gas destined for industries in the North). As expected, the Communists and another small left-wing party opposed the bill and demanded that the natural gas industry be entirely nationalized. The moment of truth was the speech of the PvdA representative Mr Nederhorst. At the end of a long, carefully prepared, and in many respects critical speech, he declared that his party would support the bill. Nothing could then stop De Pous's bill. After only one day of debate, the Second Chamber unanimously voted in favour of the bill.

Finalizing the new regime

The natural gas bill contained only the principal features of the future regime. It was now time to work out the legal and economic details of all necessary contracts and agreements. This was a crucial and complex task, and De Pous and his staff consulted the Ministries of Justice and Finance to make sure arrangements were legal and acceptable to the state. The legal departments of Shell, Esso and DSM were also very active.

It was also time to build up Gasunie, the central figure in the future gas industry. In the early Spring of 1962, Shell, Esso, DSM and the ministry informally appointed the future board of Gasunie, which would also be the board of Maatschap to ensure a close coordination between production, transport and sales. This informal board met once a week to settle important issues. One urgent task was to set up Gasunie's organization and to integrate the staff of SGB and other regional gas companies within the new company. Another important task was to start negotiations with municipal gasworks about gas prices. Two technical issues were also important: the development of the high pressure pipeline grid, and the impending conversion from town gas to natural gas.[20]

The atmosphere at the meetings of the board was very open and frank, and it seems that the board worked in the typical Dutch way of consensus building, very different from the American way: Esso's chief lawyer, who had come to the Netherlands to check all the details in the contracts and who had been invited to the board's meetings, was utterly astonished by the board's working methods. But he and the other Esso staff accepted them.[21]

The new regime at work

Building a national system

On 6 April 1963, the statutes of Gasunie were signed at an impressive ceremony at the Castle of Wassenaar, near the Hague. Gasunie symbolized the new regime, the *natural gas regime*, established in the Dutch gas industry. In this final section the properties of this new regime will be illustrated through two of the major challenges it had to cope with: the establishment of a new national gas system and the 'oil crises'. In conclusion, the role of the Groningen gas field for the emergence of a European gas network will be briefly discussed.

The establishment of a new national gas system implied two tasks. The first was to build a national grid connecting the Groningen field with all parts of the country. When Gasunie was founded, it acquired the existing regional gas grids that had been built by SGB and DSM, representing a total of almost 3,000 km. A few years later Gasunie also bought Hoogoven's regional grid in North Holland. These regional grids were insufficient as national trunklines, and thus a whole new, high capacity grid had to be built.

This turned out to be a challenge because two-thirds of the Netherlands are at, or below, sea level and criss-crossed with rivers and canals, which implies that there is a high concentration of water in the soil. Combined with the generally very thin layer of topsoil and the high population density, this made pipelaying a technically and financially demanding task (Peebles *et al.* 1971: 4f). Legal problems also had to be solved. Most important was to acquire the permissions needed for building pipelines. In January 1964, a special coordinating committee with representatives from the ministries and authorities involved in physical planning was created. Its task was to coordinate and speed up the planning process and to give final approval to the proposed rights of way.

The committee was successful: all permissions were obtained in time and many thousands of property owners were compensated (Peebles *et al.* 1971, Kielich 1988: 58f).[22] The actual construction of the new grid system started on 1 April 1964. By the end of 1969, 1,700 km of large diameter pipelines and 2,200 km of regional grid lines had been built connecting all the local gasworks with the Groningen field (Peebles *et al.* 1971: 5).

The second task implied by the establishment of a national gas system was to convert all gas appliances in the country to natural gas. The preparations for this enormous operation were carried out in close cooperation between Gasunie and the local and regional gasworks, which had formed a new organization, SROG, in 1963. It was decided to create a new company, Gascon NV, for this purpose. Gascon started operations in 1964 when the first pipelines were being built. Within four and half years the whole conversion was completed, with a peak of nearly 740,000 customers converted in 1966 (Kielich 1988: 75f, Peebles *et al.* 1971: 5).

The strategy for gas sales developed by the Esso team was carried out very purposefully. After negotiations between Gasunie and SROG new gas tariffs were introduced and a massive advertisement campaign took place. In a few years (during the second half of the 1960s) more than 60 per cent of all households connected to the gas grid decided to convert to gas heating. The average gas consumption per household increased more than five-fold. In fact, gas sales increased faster than expected in all market segments, and in 1974 natural gas had become the major energy source in the Netherlands supplying more than half of the total energy needs. The new regime thus appears to have been perfectly well suited to the establishment of a nationwide natural gas system; the objectives pursued were even surpassed.

The oil crises

In the early 1970s the gas regime was put to the test when the gas industry faced new challenges. During the 1960s, it was commonly assumed that energy prices would decrease in the long run. But in the early 1970s expectations changed dramatically. In 1972 the Club of Rome presented its report entitled *Limits to Growth* which among other things predicted that the world reserves of natural gas would be depleted within one generation if the present energy policies were not changed. This report had a strong impact in the Netherlands. Another change concerned nuclear energy. In the early 1970s the safety and profitability of nuclear power generation were increasingly questioned, and an anti-nuclear movement gradually developed. The most profound change, however, came in 1973 with war in the Middle East and the subsequent drastic rise of oil prices.

This context generated an intense debate on energy policy in the Netherlands, which in particular caused profound changes in the national gas policy from 1974, when the fast expansion policy was substituted for a no-expansion one. This policy change was implemented through the special powers granted by the natural gas bill to the Minister of Economic Affairs: Dutch gas production decreased from a peak level of almost one hundred billion cubic metres per year in the period 1976–9 to a level of about 80 billion cubic metres in the early 1990s. This demonstrates the strong governmental influence on the natural gas regime.

Another issue raised in 1974, when energy prices started to increase dramatically, concerned the allocation of the benefits raised from gas sales. The story illustrates the significance of Shell's role in the regime. Already in 1971 the allocation rule of gas profits between the State and NAM had been partially changed from 70/30 to 85/15.[23] Placed under strong political pressure to change the rule once more, the government decided to act. The Minister of Economic Affairs, R. Lubbers, agreed with the president of Shell, G.A. Wagner, to change from 85/15 to 95/5. When Wagner informed Esso's president of the agreement, the latter exclaimed 'What are you doing?', to which Wagner replied 'The best possible'.[24] As a Dutchman, Wagner knew how the country felt and understood that a concession had to be made in order to preserve the existing regime.

Thus, in the mid-1970s, the Dutch natural gas regime was able to cope with and adapt to a profoundly changing context, by modifying such crucial elements of the regime as the long-term sales strategy and the rules for the allocation of profits.

Epilogue: a new international regime

The discovery of the Groningen field affected not only the Dutch energy system, but also neighbouring countries and, to a certain extent, the entire European energy market. Shell and Esso played a crucial role in this process. They took an active part in the creation of gas grid companies in West Germany and Belgium which built the pipelines needed to export Groningen gas. As early as 1964, exports to Germany began, in 1966 to Belgium and in 1967 to France. Since the late 1960s, between 30 and 40 per cent of the gas produced in the country was exported, and for many years the Netherlands was the largest gas exporter in Europe. The exploitation of the Groningen field thus stimulated the building of cross-border pipelines that eventually turned into an integrated European gas network (Davis 1984, ch. 8, Estrada *et al.* 1988, chs 2 and 8, Lönnroth 1985: 4).

As for domestic sales, maximum profits were sought by Gasunie. The prices in export contracts were therefore set close to the prices of alternative fuels, primarily heating oil. Shell and Esso, through their subsidiaries in many countries, had a unique overview of the European energy markets, which was of great importance when negotiating these contracts. It is interesting to note that, as in the Netherlands, this pricing policy often coincided with the interests of the major gas buyers. Many of these had interests tied to domestic energy supply industries, such as the German coal mines and the French and Italian domestic natural gas fields. They were happy with gas prices that did not upset existing energy markets (Davis 1984: 182, Estrada *et al.* 1988: 9f).

The export contracts signed in the mid-1960s became a model for subsequent contracts. They shaped the legal framework for the cross-border gas trade and also influenced the more informal 'rules of the game' in the European gas industry. Thus players in the Dutch natural gas industry had an important role in shaping a new gas regime at a European level too. However, analysing the emergence of this international regime and its subsequent evolution is another story, well beyond the scope of this chapter.

Acknowledgements

This research was supported by a grant from the Swedish council for the planning and coordination of research (FRN). A number of colleagues in the Netherlands and Sweden have provided constructive comments on a draft of the chapter. I would especially like to thank Adri de la Bruhèze, Olivier Courtard, Mats Fridlund, Thomas Kaiserfeld, Karel Mulder, Marie Nisser, Ulrika Sax, Hans Schippers, Jane Summerton, Joar Tiberg, Geert Verbong, Onno de Wit and Hans Weinberger.

Notes

1 'It was a bonanza!', was the expression Mr Wagner, former president of the Royal Dutch/Shell Group used when he referred to the discovery of the Groningen field. Interview with G.A. Wagner 15 April 1994.

2 This development towards municipal dominance is similar to what happened at the same time in the Nordic countries (Hyldtoft 1995).

3 The so called Wobbe-index defines the burning properties of a gas. This index is defined as the quotient between the energy content and the square root of the density of the gas in relation to air. Two gases that have the same Wobbe-index can be used in the same burner even if they have very different chemical compositions.

4 There is a vast literature on the pillarization of Dutch society. A good overview is given in Jamison *et al.* 1990: 122f.

5 This law was enacted in 1810 during the short Napoleonic period. It states that underground minerals do not belong to the landowner but to the Dutch State, and exploitation is subject to state concession.

6 Wansink wrote a great number of internal memoranda to the Minister concerning the important issues throughout the process. These memoranda reflect the reasoning of the Minister and his staff. They are kept at the Semi-statisch Archief van het Ministerie van Economische Zaken, Archief van het Directoraat-Generaal voor de Energievoorziening: Gasvoorziening, boxes 104 and 155. In the following I shall refer to these archives as the 'MEZ archives'.

7 Shell investigated to check whether the 1932 agreement was still binding. Only afterwards did Shell agree to form NAM with Esso. Interview with Wagner (Shell's investigator), 15 April 1994.

8 Interview with J.P. van den Berg, one of the members of the Esso team and former Sales Manager in Gasunie, 20 April 1994. See also Kielich 1988: 32f.

9 I have not found the report of the ESSO-team at the MEZ archives. The figures are from Peebles *et al.* 1971: 19.

10 Interview with van den Berg, 20 April 1994. See also Kielich 1988: 32f and Peebles *et al.* 1971.

11 MEZ archives, box 155.

12 Interview with L.G. Wansink 25 April 1994. See also *Nota inzake het aardgas* 1962: 10.

13 In MEZ archives, box 155, one can see how these arguments are gradually developed, in internal memoranda primarily, during the spring of 1961. They are summarized in Nota 1962: 4f.

14 For example, SGB outlined their views to the Minister at two meetings on 31 October 1960 and on 21 August 1961 (MEZ archives, boxes 104 and 155).

15 MEZ archives, box 155, my translation.

16 Wansink called it a 'strategic mistake' that there had not been a representative from the Ministry present. Interview with Wansink 25 April 1994.

17 I have not found the report in the MEZ archives. The main points of the report are outlined in Nota 1962: 5f; see also Kielich 1988: 39f and Salverda and Bakker 1983: 18f.

18 Ibid. MEZ archives, box 157, contains a copy of a letter dated 11 December 1964, from the executives of Gasunie and NAM/Gas Export to the board of the Gasunie, stating explicitly that the only function of NAM/Gas Export is to send invoices to foreign customers. Payments were to be transferred immediately to Gasunie. In 1975, when many OPEC countries had nationalized their oil industries, this façade was no longer needed and NAM/Gas export was simply abolished.

19 Interview with Wansink, 25 April 1994.

20 Ibid. See also Kielich 1988: 49f.

21 Interview with Wansink, 25 April 1994. See also van Iterson and Olie 1992.

22 Mr Wansink remarked that if Slochteren had been found ten years later, there would have been much more opposition to pipeline building from environmental and other groups (interview with Wansink, 25 April 1994).

23 Profits were split 70/30 up to a given price level, and 85/15 above that level.

24 Interview with Wagner 15 April 1994. See also Davis 1984: 161f.

References

Borghuis, J.B. (1988) *Veertig Jaar NAM. De geschiedenis van de Nederlandse Aardolie Maatschappij 1947–1987*, Assen (NL): NAM B.V.

Davis, J.D. (1984) *Blue Gold: The Political Economy of Natural Gas*, London: George Allen & Unwin.

Dil, G. and Homburg, E. (1993) 'Gas', in H.W. Lintsen, M.S.C. Bakker, E. Homburg, D. van Lent, J.W. Schot and G.P.J. Verbong (eds) *Geschiedenis van de techniek in Nederland. De wording van een moderne samenleving 1800–1890*, Part 1, Zutphen (NL): Walburg

Estrada, J., Bergesen, H.O., Moe, A. and Sydnes, A.K. (1988) *Natural Gas in Europe. Markets. Organization and Politics*, London: Pinter.

Falkus, M.E. (1967) 'The British gas industry before 1850', *Economic History Review* 10, 3: 494–508.

Homburg, E. (1986) 'The history of the Dutch chemical industry', *Chemical Magazine*, special issue: 16–22.

Hughes, T.P. (1983) *Networks of Power: Electrification in Western Society 1880–1930*, Baltimore (MD): Johns Hopkins University Press.

Hyldtoft, O. (1995) 'Making gas: the establishment of the nordic gas systems 1800–1870', in A. Kaijser and M. Hedin (eds) *Nordic Energy Systems. Historical Perspectives and Current Issues*, Canton (USA): Science History Publications.

Iterson, A. van and Olie, R. (1992) 'European business systems: the Dutch case', in R. Whitley (ed.) *European Business Systems. Firms and Markets in their National Contexts*, London: Sage Publications.

Jamison, A., Eyerman, R. and Cramer, J. (1990) *The Making of the New Environmental Consciousness. A Comparative Study of the Environmental Movements in Sweden, Denmark and the Netherlands*, Edinburgh: Edinburgh University Press.

Kaijser, A. (1987) 'From local networks to national systems: a comparison of the emergence of electricity and telephony in Sweden', in F. Cardot (ed.) *Un siècle d'électricité dans le monde*, Paris: Presses Universitaires de France.

—— (1992) 'Obstacles to international collaboration: the emergence of the Nordic power exchange', in M. Trédé (ed.) *Electricité et électrification dans le monde*, Paris: Presses Universitaires de France.

—— (1995) 'From invention to global system', in M. Karlsson and L. Sturesson (eds) *The World's Largest Machine. Global Telecommunications and the Human Condition*, Stockholm: Almqvist & Wiksell International.

Kielich, W. (1988) *Ondergronds rijk: 25 jaar Gasunie en aardgas*, Groningen (NL): N.V. Nederlandse Gasunie.

Lönnroth, M. (1985) *Troll Dance. The Next Act on the West European Gas Scene*, Stockholm: Beijer Institute.

Messing, F.A.M. (1988) *Geschiedenis van de mijnsluiting in Limburg*, Leiden: Martinus Nijhoff.

van der Noort J. (1993) *Licht op het GEB. Geschiedenis van het Gemeente-Energiebedrijf Rotterdam*, Rotterdam.

Nota inzake de gasvoorziening, aangeboden aan de Tweede Kamer der Staten-Generaal (1953), NL Door de Minister van Economische Zaken.

Tweede Nota inzake de gasvoorziening (1958), NL Zitting 1956–1957: 4900.

Nota inzake het aardgas (1962), NL Zitting 1961–1962: 6767.

Peebles, M.W.H., Pass, V.H. and Salkeld, J. (1971) *The Development of Groningen Gas. A Short History of the Development of Groningen Gas Up to End 1969*, Report No. SIG 71/6, Shell International Gas Limited.

Salverda, F. and Bakker, V. (1983) 'De jacht op ons aardgas 1', *Nederland*, 15 Jan.: 3–36.

Thue, L. (1995) 'Electricity rules. The formation and development of Nordic electricity regimes', in A. Kaijser and M. Hedin (eds) *Nordic Energy Systems. Historical Perspectives and Current Issues*, Canton (USA): Science History Publications.

Yergin, D. (1991) *The Prize. The Epic Quest for Oil, Money and Power*, London: Simon & Schuster.

4

TECHNICAL SYSTEMS AND STRATEGY

Intercontinental telecommunications in the first quarter of the twentieth century

Pascal Griset

> The ease one feels when formulating a project never returns to the same degree when one is actually involved in executing it. When calculating the chances of success, there is no risk, whereas, once one is involved in the execution, fear appears, depriving us of our capacities.
>
> Thucydides, *The Peloponnesian War*

During the second half of the nineteenth century, intercontinental communications had been exclusively based on cable technology. The advent of radio at first hardly challenged this position. Yet, by the end of the First World War the two technologies had evolved at such different rates that the status quo was no longer tenable: whereas wireless telegraphy had undergone rapid change, submarine cables technologies had only experienced modest improvements. Yet, does it imply that as soon as the early 1920s radio represented a technical tool coherent and efficient enough to threaten the domination of cable? Can it be said that telecommunications had already moved from a single to a dual technology?

We will explore these important issues in three ways. First we will examine the extent to which radio challenged cable in that period, keeping in mind that judgements then made were based as much on subjective feelings as on rational reasoning. Then we will discuss the impact of technological evolution in telecommunications on the balance of geostrategical powers. Finally we will look at the specific action of private companies and their new role in the expression of a national ambition.

Cable challenged by radio?

Around the middle of the nineteenth century, Great Britain began to develop an underwater cable network which aimed at establishing links between London and all British colonies. With the exception of transatlantic links with the United States, this network was set up as part of an essentially strategic plan to use Britain's resources, to assure the country's security and at the same time to support its expansionist policy. Nearly 250,000 km of intercontinental cables were controlled by British firms by the end of the century. In 1896, Marconi performed the first wireless telegraphic transmission. Radio quickly appeared as a revolutionary technology, and a threat to the existing cable system. Yet, its enormous potential was still waiting to be exploited.

What was the state of the cable system at this time? While this system had been substantially modernized during the last years of the nineteenth century, prices to users had stopped decreasing after the strongly competitive period of the 1880s. 'Transmission is much faster, operating costs are lower than they were fifteen years ago, and yet we are still paying the old rates [. . .] We live in an era where prices are lower and lower – except for cable,' wrote an indignant *New York Times* in 1899 (15 August 1899: 6). In 1902, the *Times* continued: 'It is evident that claims to the effect that the transatlantic cable system is already running at full capacity and any reduction in rates would be ruinous, are incompatible with the arithmetic' (9 March 1902: 6).

Pricing policy wasn't the only bone of contention. Despite their commercial success, cable companies seemed incapable of modernizing their system: 'Western Union and the Postal Telegraph Company use archaic systems,' the *New York Times* opined in 1902; 'to our knowledge, Western Union spends no more than 10¢ per year on development. It's as if there were a huge sign over their door: "Inventors and Scientists Not Wanted"' (1 May 1902: 8). 'The cable companies are about as incapable of improvement as the channels on planet Mars, and they are managed with about as much concern for the wants and needs of the inhabitants of Earth,' raged the *Times* a few years later (26 September 1907: 2).

After the grand manœuvres of the 1880s and 1890s, the system seemed paralysed. As late as 1921, the *Annales des PTT* were still denying that the future of submarine cables might be 'really compromised' (Annales des PTT 1921); even to them, however, the problem seemed worrying. Under criticism, providing lower-quality service, the transatlantic cable companies seemed a long way from the smugness of the 1890s.

Those who were disillusioned with cable placed their hopes in radio. Even before the first transatlantic radio connection had been established, there was the hope of an alternative based on wireless telegraphy. In 1899, the *New York Times* announced: 'competition from wireless telegraphy will give rise to wonderful price reductions, and result in its being used in ways still inconceivable in everyday business' (7 May 1899: 20). Reactions in 1901 to

Marconi's announcement of the first wireless transmission across the Atlantic are suitably eloquent: although his achievement was more of an exploit than a commercial first, Marconi was sued by the Anglo American Telegraph Company, which argued that this constituted a breach of its monopoly in Newfoundland. In private, Marconi was delighted with Anglo American's disproportionate reaction, calling it 'the greatest tribute to our success' (quoted in Baker 1902).

These events resulted in a substantial fall in stock prices of several cable companies, with Anglo American the hardest hit (*New York Times* 22 December 1901: 8). G. Marconi had pulled off a coup in the eyes of the public. He offered an alternative, playing on the new technology's dynamism and novelty in the face of cable companies often caricatured (e.g. in the press) as dangerous snakes or pervasive octopuses.

In a few years, this negative impression of the cable companies became deeply rooted in public opinion. The American public had one of those clear cases of heroes versus villains which they enjoyed so much. As S. Douglas pointed out, 'The press had created a romanticised and sentimentalised preference for wireless telegraphy . . . the cable companies were the monopoly that the public loved to hate' (Douglas 1979: 236–7). It must be noted that if the press was indeed fulfilling its role in informing the public of new technological developments, it also had a vested interest in putting pressure on the cable companies to lower their rates.

But could the available radio systems live up to the expectations placed in them? Were they commercially viable? The answer to this question was clearly negative before the First World War, but things were far more complex by the beginning of the 1920s. Numerous studies were conducted. One of the most exhaustive was Alexanderson's in New Brunswick, which analysed a radio station emitting twenty words per minute (Alexanderson 1919). A similar study was carried out in France in 1922; its conclusions concurred with Alexanderson's. Based on reasonable assumptions, the French experts estimated that radiotelegraphic stations would be able to compete with submarine cables (Chaulard 1922).

Serious reservations were none the less expressed about the hertzian airwaves' ability to replace submarine cable connections. Even the *New York Times*, which had always been a staunch supporter of wireless telegraphy conceded in 1919 that radio was 'still in its infancy' (23 November 1919), reflecting the then prevailing view. From the customer's point of view, there were still important problems with radio as compared to cable. The lack of secrecy was undoubtedly the most frequent and the most valid criticism. The Associated Press did not favour radio connections at all. The news AP transmitted to its customers could just as easily be picked up by publications that were not paying for the service. As AP's general manager, Frederick R. Martin, declared in 1920: 'we have no prejudice against radio, we simply wish to express certain objections that radio inspires' (quoted in US Congress 1921: 284).

The data available in the early 1920s thus allowed a precise comparison of submarine cables and long wave radio transmission. High-frequency alternators made radiotelegraphy quite competitive. For approximately the same investment as that required for a cable (about 60 million francs), a high-power transmitting station had the advantage of allowing multiple connections. Moreover, these connections could easily be changed – whereas a cable could not be altered once it had been laid. But if radio stations compared favourably to cable in terms of investment costs, their operation required costly supplies of energy, and radio connections were not as good as cable connections. Although not as clearcut as the radio industry might have wished, the situation was nevertheless much more favourable to radio than a few years before.

The available technical potential was hence particularly difficult to assess in the period immediately following the war. Two main options were available to anyone wishing to open a transatlantic connection: submarine cables or radio. In 1902, the London *Electrician* had already surmised that if submarine cables had been discovered after radio, 'what enthusiasm this new invention would have generated throughout the world. Cable, which transmits messages directly to their destination without dispersing them in all directions – what a marvellous advance compared to hertzian radio waves' (*The Electrician* 17 January 1902: 501).

The same comment could have held true twenty years later. Though its position was better than it had been in 1919, radio remained, owing to its operating costs, the lack of secrecy in its transmissions and the inconsistency of most radio connections, a mere complement to intercontinental cable. Though radio technology seemed to open the way for a new industry which would overtake several decades of investment in submarine cables and give a second chance to latecomers, cable remained the best all-round solution, combining secrecy and reliability. 'In the years to come, cable will remain a vital component of the worldwide communications system. No country will be able to afford to be without either radio or cable,' affirmed a US Navy report, 'a race has begun between the US, England, France and Japan for control of the world' (US Navy 1919). The radioelectric industry could still not claim to be based on a coherent technical system. Divided between its electrical origins and its electronic future, aware of its potential but unable to materialize it in the short term, made up of young companies the radioelectric industry had no technological certainties on which to project its future. Intensely committed to the development of electronic components, it was still entirely dependent on the most classical electrical technology when building its intercontinental transmitting stations. The choice between radio and cable still could not be made according to financial and technical absolutes. Hence, the early 1920s were, for transatlantic connections, truly an intervening period between two technologies.

Upsetting the balance of geostrategical power

Until the beginning of the twentieth century, intercontinental telecommunications were based on a relatively coherent technical system which allowed for clearcut decisions. Objectives to be pursued were well-defined, results obtained were visible. Also, the management of international telecommunications was stable and efficient, and carried out by well-established institutions. The advent of wireless telecommunications, however, upset this equilibrium.

Competition from American cable companies, and secondarily from French and German ones, considerably weakened the dominant position of Great Britain. Although their supremacy was still assured on a global level, the British had more and more difficulty holding on to the most profitable market, the North Atlantic. Formerly a ground-breaking, economically profitable and politically effective vanguard industry, British cable companies had weakened and now needed to be protected. Their contribution to the influence and to the very functioning of the empire necessitated public intervention, even if liberal ideology had to suffer. K.C. Underdown, chairman of the Direct United States Company considered in 1901 that capital invested in the British cable companies should to some extent be protected. He explained: 'We regret seeing our heritage diminish regularly in the last few years due to the continual installation of cables owned and operated by foreign countries' (Underdown 1902: 51). Up to then the motor behind the development of cable companies, liberalism, had become undesirable and was analysed as a factor contributing to their decline. The excessive freedom allowed foreign companies was the focus of criticism. Because of this, a new examination of cable activity was undoubtedly called for.

Through a lack of strong companies, Britain was forced to separate economic and commercial stakes from strategic stakes. Giving up the worldwide strategy it no longer had the means to maintain, the British government hierarchized its intercontinental system. Links with the empire were the top priorities: each territory should be connected to London by a cable completely independent of foreign control. Other links were no longer considered as priorities in the country's cable policy. They were considered only in commercial terms. It was in the framework of this new doctrine that the leasing of the entire British northern transatlantic cable system to Western Union was decided upon. Hence, in 1912, not a single transatlantic cable remained under the economic control of Great Britain.

These fundamental changes in the economy of the transatlantic network were still not enough in the eyes of the Americans. In any case, their victory was spoiled by the arrival of a newcomer: Marconi Wireless. Soon after the First World War, the American government was becoming worried about an eventual return of British domination of intercontinental link-ups. In 1918, a source within the US Navy exclaimed with alarm: 'It is generally recognized that the Marconi Company benefits from the unofficial support of the British government, and the history of that company demonstrates that, although unofficial, this support is none the less strong. There seems to be no doubt that that company has a well-established plan to control international commercial radiocommunications' (US Navy 1918: 11). And a few years later:

> This is, in fact, the beginning of a plan which aims to cover the entire planet with a network of purely British radio transmitting stations. It is foreseeable that, when an international conference is called, Great Britain will ask participating countries to recognise a large series of radio frequencies which it will claim are necessary for the proper functioning of its network. The nature of this list is already known.
>
> (Secane 1923)

To a lesser degree, the activities of other European powers were also being watched with concern: 'Today the European nations (Germany, England, France) are covering the world's airwaves with radio news programmes. This is purely and simply government propaganda', commented United Press International with indignation (quoted in US Congress 1921: 181). Radio was considered as much a danger as a formidable opportunity. The US Navy diagnosed:

> The apparently insurmountable difficulty the United States are having in developing a real system of submarine cables has become a stumbling block that is preventing any partial improvement of the existing network. Fortunately, these problems don't exist in the domain of radio. While the potential for manufacturing submarine cables is concentrated in Great Britain and the sources of the necessary raw materials are firmly in British hands, the situation is the reverse for radio. The biggest and most dynamic manufacturers of radio equipment are American, and the patents they control are reputed to be superior to those held by foreign companies.
>
> (US Navy 1918: 11)

A real collaboration between private enterprises and the government could thus constitute a major asset for the United States:

> If our government could coordinate the activities of private companies with those of government infrastructure, we would find ourselves in a much better position than France or Great Britain which are going in the direction of pure state control. These governments would see their activity limited to their own territory. The

> hostility that exists in certain European countries between public and private radio systems is therefore to our advantage if we have enough foresight to be able to take advantage of it.
>
> (Bryant 1923)

While the Americans divided their attention equally between radio and cable, Great Britain, unable to break with its past, wasted time and missed opportunities. France and Germany, on the contrary, saw in radio a chance to make up for lost time. Britain's inability to assimilate and apply the lucid 1910 analysis of the real value of submarine cables prevented it from fully profiting from the asset Marconi developed. The Americans were intrigued by this attitude. A 1919 US Navy memorandum estimated that Britain was in the best position to develop a world network of radiocommunications rapidly. And yet, the memo concluded, 'the British government has asked the cable companies to expand their system and develop their factories' (US Navy 1919).

The symbolic and financial weight of cable companies hardly favoured the development of international radio-electric networks based in London. Though Marconi flourished, it did so in an unfavourable environment. Great Britain's policy, while not totally neglecting wireless telegraphy, remained essentially cable-based until the mid-1920s.

Due to the situation of its public institutions, France was undeniably close to the British schema. France also had a powerful public monopoly, undoubtedly even more rigid than the British one. Also, its political culture, its Empire and a certain vision of how its place in the world should be expressed pushed France in the direction that Britain had taken. However, as the American analysis emphasized, unlike Great Britain France did not have the advantage of a cable system on a par with her ambitions. 'France has always been insufficiently equipped with submarine cables,' commented a diplomat in 1922. The initiatives taken at the beginning of the century did not give rise to any long-term plan of action. Articles appeared in the French press that could easily have been written twenty years earlier. Two articles published in *Questions Diplomatiques et Coloniales* summarized a long list of complaints:

> London has quite rightly been called the central market for world news, but one would have to have a very incomplete understanding of the British domination – or, more precisely, monopoly – if one imagined that the English would pass it on without profiting from it, or that they would re-sell immediately the news to which they had had first access. Their industrialists, their businessmen, their financiers have the great advantage of being informed before their foreign competitors.
>
> (Jacob 1922)

Of course, as we have already indicated, the PTT (the French public telecommunications administration) continued to favour the protection of existing cables at the expense of radiotelegraphy. At the same time, the counterweight of the older technology was too weak to hold back the development of radio after the First World War. France took advantage of this opportunity to try and make up within a few months for a gap that had accumulated over several decades. After being totally dominated by the British and by the Americans since the mid-nineteenth century, France quite rapidly hit upon a solution that was both technically brilliant and financially manageable.

This new-found status was noted in foreign communiqués: 'Since the war, France has turned its attention toward the development of wireless connections, especially for its colonies. Consequently, France has built transmitting stations that are among the most powerful in the world' (US DOS 1922). Sainte-Assise was for a few months 'the most powerful station in the world'. Even more revealing, the French CSF was in a position to deal with RCA on an equal footing when negotiating traffic agreements, and was able to obtain a very favourable agreement concerning patents. The pragmatic attitude supported by Briand became a doctrine for French telecommunications.

The new enterprises: driving forces of change

In direct contrast to the indecisiveness of governments and administrations disorientated by the barely conceivable changes about to take place, a formidable force was emerging: the radio-electric companies. Newly founded, often by brilliant and original minds seeking risk and profit and aware that lethargy was synonymous with creeping death, these companies were the driving forces of change.

The key people in the radio-electric companies were the engineers. Their motivation was based primarily on their passion for this new technology which they sought to improve from one month to the next. This was the spirit that the first generation of cable companies had had. The Cyrus Fields, the Siemens of the radio era were driven by their desire to push a radically new technology ever further. Their success rested on their ability to make radio evolve faster than their competitors.

These men were certainly not philanthropists; they were aware that they were part of a profit-based system and in no way neglected the financial aspect of their activity. Yet profit remained in many ways secondary to them. Money seemed to represent more a confirmation of success than a goal in itself. Rather than 'pioneers', a term which seems to lend itself naturally to such individuals, we see them as 'system builders' (a term coined by Thomas P. Hughes). Marconi, Sarnoff and Girardeau, unlike De Forest, were able to occupy the economic terrain, and although they had certain technical knowledge they weren't strictly speaking technicians. Sarnoff was merely an operator; Girardeau an engineer from the prestigious Ecole Polytechnique who never contributed anything significant in terms of research. Marconi himself, after opening up a new domain largely by exploiting the advances made by others, surrounded himself with top people; his role was essentially to supervise their work (often brilliantly). Without being great technical geniuses they based their action on technology, their hopes rooted in an unshakeable faith in progress and in their love for radio.

This sense of a mission can be found in the writings of Sarnoff. His analysis of the technical climate just after the war shows how radical the break was with the mentality of cable systems' managers. It completes our demonstration and illustrates an important characteristic of these men:

> Given that cable has been operative for more than fifty years, it seems almost unbelievable that today, the fastest transatlantic cable cannot transmit more than an average of 150 to 200 letters per minute. Instead of developing the speed of their cables, which would have been the desirable and scientific attitude to take, the cable companies were content to lay more cables. The companies were too quick to resign themselves to the limitations of their lines, they did not maintain competent research laboratories.
>
> (Sarnoff 1920: 6)

This then was the 'anti-model' depicted by Sarnoff. His own conception of the evolution of communications was quite different:

> On the other hand, radio communications have experienced more scientific and technical progress in the last five years than the cable companies throughout their whole existence. Now that radio is in its commercial phase, I envisage the next five years as being even more prolific in inventions and improvements. Our company should be following an ambitious policy based on research, invention and development.
>
> (Sarnoff 1920: 6)

A similar technical culture prevailed at CSF, the less powerful structure headed by E. Girardeau. Moreover, radio companies had a very modern conception of society. They adapted their technologies to social needs and expectations. This role of go-between translated itself into a dynamic commercial strategy which consisted of rapidly attacking new markets, and even facilitating their emergence through the offer of products and services.

According to reports on the development of radio in the early 1920s it would have seemed logical to wait for a while before setting up an intercontinental radio-electric system. Such an investment would have been difficult to integrate into the strategy of young growing companies concerned with their competitors and anxious to find outlets as quickly as possible.

Radio understood that it could not compete directly with cable and chose to supply complementary, rather than competitive, products.

Competition with cable also occurred in discourses. As we have pointed out, submarine cable systems supported an imaginary world order quite vivid in the political and economic debate. After the war, radio in turn provided for a political metaphor. While cable was part of an image based on control – a network covering space little by little – radio suggested the power of the monumental. Where cable exercised discreet, invisible power, radio needed gigantic, spectacular installations. Long wave transmitting stations installed at the beginning of the 1920s were of monumental proportions.

The antennae of Radio France at Sainte-Assise were supported by twelve pylons, each 250 metres high. This was not unique to French technology. Rocky Point, the transmitting station of RCA's 'Radio Central', received two groups of antennæ at the beginning of 1922. Twelve had been planned. The two groups already installed were supported by six steel pylons, each 125 metres high and weighing 152 tons; the whole group supported by concrete foundations that weighed 8,250 tons. The antenna itself was made up of 80 km of silicon bronze, and the ground connection used 725 km of copper wire. The electrical system was attached to the network by a 23,000 volt cable (Annales des PTT 1922). This was the general tendancy. Initiated by Marconi during the first decade of the century, it became the norm with all the radio companies (Vasseur 1975: 105).

The desire to transmit on longer and longer waves led engineers to design more and more enormous antenna systems and extremely high power transmitters. The Dutch Telecommunications Service pushed this to the limit with its station at Bandoeng in the Dutch West Indies. Built in 1923, it became the most powerful in the world with a 2,400 kW oscillating arc. The magnetic circuit alone weighed 260 tons. The layered antenna was so colossal that it was thought wise to forget about pylons; instead the antenna was strung between the two slopes of a valley.

Rather than merely being tolerated as a necessary evil, the enormity of these stations was actually an aspiration of their designers. On advertising billboards and in articles in the press, the words 'power' and 'gigantic' reappeared frequently, and the unofficial title of 'the most powerful station in the world' was claimed in turn by more and more colossal installations forming a sort of 'blue ribbon' drawn by the various stages of a race among nations for the control of the ether. These installations had a double 'personality'. Financially more accessible than submarine cables, they made it possible to establish entire networks in a short time, whereas it would have taken years with cable technology. At the same time, their considerable cost and their mysterious appearance (in the eyes of the ordinary citizen) gave them an aura of prestige.

The psychological dimension of this first generation of transatlantic transmitting stations supported their commercial impact. Although the only truly valid argument for radio as opposed to cable was economic, long wave stations, which had actually become more and more costly, were 'sold' for their prestige. From the poor man's telecommunications tool, they became a symbol of ostentation. In the spring of 1921, RCA organized a demonstration of its New Brunswick installation. Einstein was the most famous guest invited to this event (RCA 1921). On this occasion, Young mentioned the construction of a new station on Long Island, emphasizing its colossal size and power:

> When this station is finished, it will be the biggest and the most powerful in the world. To give you an idea of the dimension of this project, I can tell you that we had to acquire approximately ten square miles of land from one owner, for a price of over half a million dollars for the real estate alone.
>
> (US Congress 1921: 329)

A. Goldsmith also evoked its enormous size in lyrical terms:

> The 72 pylons of the central radiotelegraphic transmitting station of New York stretch out the equivalent of about one half the length of Long Island. [. . .] the human eye cannot see the radiowaves radiating from this powerful station. If they were visible, the pulsating dots and dashes sent out by the twelve antennæ would produce the same effect on the eye of the observer as if the whole installation were consumed in an enormous fire.
>
> (Goldsmith 1921)

Intercontinental stations had become objects of prestige for nations with global aspirations. Foreign delegations were taken on tours of them during their visits. Thus Sainte-Assise was visited by numerous delegations during its first years of existence.

The choice of long wave was not only prestigious, it was also perfectly adapted to another purpose, that of centralization. Originally welcomed as a tool of emancipation for countries and companies dependent on the powerful cable companies, radio had not fulfilled this hope twenty years later. Instead of being the poor man's telecommunications tool, it in fact became a toy for the *nouveau riche*. Supposedly a tool of decentralization, it became a new mechanism of control over the dependent countries. The choice of long wave actually gave rise to a mentality of centralization. As Alexanderson explained:

> The economic factors that lead to adopting a central radio transmitting station as the concrete solution to long distance communications are practically the same as those that led to the creation of central

> electrical energy stations . . . both assured the optimum utilisation of capital.
>
> (Alexanderson 1921)

From this infallible logic a multitude of projects were generated, all using a centralized transmitting station to handle the entire traffic. The technicians (sellers) were in perfect harmony with the vision of the politicians (buyers). The implicit politicians' demand for centralization was in this way somehow justified, i.e. the choice of a centralized station being 'imposed for technical reasons'. In this way, the political will to control the flow of information accorded perfectly with the commercial priority of creating a convergence of radio traffic, and hence of transmission taxes, into the investor country. Goldsmith for RCA supported this logic, as his 1921 statement shows:

> Certain countries, the United States for example, are, because of their geographic position, perfectly situated to act as the relay point for intercontinental communications. Communications between Europe on the one hand, South America and the Far East on the other, go through the United States. The energy necessary to transmit over a certain distance increases by a lot as the distance increases. This is why it is advantageous to have relay points that will receive messages bound for South America and the Far East, and re-transmit them to these countries, automatically or otherwise.
>
> (Goldsmith 1921)

In fact, the radio companies were marketing a communications tool that used exactly the same geostrategic system as cable. It created a centre with its periphery, it suggested the expansion of the nation's influence and thus was perfectly accepted by a political class that had been frustrated for decades by British domination. The so-called blindness of radio-electric companies regarding short wave technology must thus be reassessed. Long wave stations corresponded to the expectations of clients. The substantial investments made by the operating subsidiaries of the large groups provided the necessary funds for manufacturing and research activities. The large sums invested also justified the relatively high rates charged, hardly less than cable rates. The construction of long wave stations thus appears as the result of a tacit agreement among various partners on a technological choice which was costly and risky, but which satisfied the needs of all the actors involved.

Conclusion

The period from 1918 to the mid-1920s seems to be a key period for understanding the evolution of large telecommunications systems in a dynamic context. The new data offered to decision makers by technology was numerous, complex and appeared heterogeneous. As we have seen, there was not one clear technical solution. Moreover, the available technologies belonged to radically different systems, technologically as well as politically and culturally. Only an unstable and fragile system could possibly emerge from this context. Geostrategic equilibrium is as much a product of the objective reality of the powers involved as of their perceptions of what their respective potential is. Between 1900 and 1920, the gap between this reality and the perceptions of the politicians appears to have been substantial. Tensions were resolved at crucial points in time when real choices could no longer be put off in favour of cosmetic modifications of the new technologies within old analytical frameworks.

This process, carried out with the tacit (not to say unconscious) agreement of the political establishment, was sustained by the commercial policies of the four major radio companies. The ability of the radio-electric companies to influence important decisions introduced a new element into the economic–strategic process. More and more vigorously, the radio companies would design their strategy themselves. Because of the vertical integration of manufacturers and operators within the same companies, unlike the prevailing conditions in the cable industry, the strategy would be focused on research and manufacturing centres. The airwaves were no longer the principal objective of these companies. Being operators above all permitted them to test their equipment, to market it and to keep profits flowing toward the shareholders of the company.

On a more global level, this reflection on large telecommunications systems could be subsumed in the concept of 'technical–strategic paradigms'. The history of transatlantic telecommunications demonstrated the intimate connection between technology and geostrategy. The United States succeeded Great Britain as the dominant power in the Atlantic, but the two countries dealt with space and used technology to affirm their leadership in different ways.

The British approach was based on a systematic and physical control of space. In a way this represented a two-dimensional vision of geostrategic space involving the control of islands, strategic points and straits, the patient weaving of a cable network, year after year extending an ever more dense net over planetary space. The time dimension was marginal. Time was perceived as linear and the system of telegraphic cables seemed frozen in time. The only capacity for developing its potential resided in a progressive extension of its infrastructure.

This conception of geostrategic superiority did include technology, but in a manner resembling a war of position where superiority could only be maintained by ceaselessly increasing one's fortifications. Like a citadel accumulating reserves and reinforcing its walls, the British cable system accumulated important funds and expanded its cables like new lines of defence. Through this static conception of its power, however, Great Britain exposed itself to successive attacks which, incidentally, could be based on similar principles. This attitude was founded on the faith in a privileged situation deriving from the localization of the British Isles, the size of the British Empire and the technological advance acquired through early investment. It seemed to guarantee the country's eternal supremacy.

The appearance of radio upset this lovely order by completely changing the space-time relationship. Geography became secondary: how important were islands, strategic points and straits when radio could cross borders and distances? True, for a few years, networks continued to be designed in the old fashion, but that did not last long. Hierarchical systems fell apart once and for all with the advent of short wave. It was no longer a question of occupying the territory physically, but of conquering high frequency hertzian space through scientific and technical means. This introduced a temporal dimension that was no longer linear, but variable. With technological advances came the speeding up or slowing down of time. Innovators could master these variations and in a way become the masters of time.

Thus a 'modern' techno-strategic paradigm was shaped, based on the new approach the Americans had developed toward technology. In this approach, the patent became as important an arm as a destroyer or the control of 'strategic points'. This understanding of innovation was the key to a dynamic vision of technological superiority. The Americans seem to have understood first that technological superiority was fragile, and depended on unremitting activity. They had to maintain a rapid pace of innovation, destined to eliminate competitors by 'suffocation'. This is how they engineered the triumph of their conception of how hertzian space should be managed: refusing *a priori* any dividing up, they demanded unhampered access to radio frequencies at a velocity dictated by new technological systems. 'First come, first served', said the Europeans; 'All power to dynamism and creativity', retorted the Americans.

The new techno-strategic order was thus founded on radically different principles than those underlying the British cable system. A different vision succeeded that of control of space by physical occupation of strategic points. A war of movement succeeded the war of position, and the new war was fought on different ground. From then on technology moulded space rather than being moulded by it. The modern techno-strategic paradigm defined a new space-time relationship, the key to which was technology.

References

Alexanderson, E. (1919) to R. Weagant (24 March), Museum of American History, Clark collection, Box 65.

—— (1921) Conference (held in October 1920), *Annales des PTT*: 364–7.

Annales des PTT (1921) 'L'avenir des câbles sous-marins est-il réellement compromis?', *Annales des PTT*, 244–5.

—— (1922) 'Le central radiotélégraphique de New York' (translated from *Radio Review*, January 1922), *Annales des PTT*: 480–3.
Baker, R. (1902) 'Marconi's achievement', *MacClure's Magazine*, February: 294.
Bryant, S. (US Navy captain) (1923) to rear admiral W. Rodgers (February), National Archives, RG 45, Box 983.
Chaulard, L. (Cdt) (1922) 'Les grandes stations radiotélégraphiques au point de vue financier', *Annales des PTT*: 361–81.
Douglas, S. (1979) *Exploring Pathways in the Ether: The Formative Years of radio in America, 1896–1912*, unpublished PhD dissertation, Brown University.
Goldsmith, A. (1921) 'World communications' (conference held at a meeting of the Institute of Electrical Engineers in New York on 17 November 1921), *Journal of the American Institute of Electrical Engineers*, December.
Jacob, L. (1922) 'Les intérêts français et les relations télégraphiques internationales', *Questions diplomatiques et coloniales*, 1 and 16 August 1922.
RCA (1921) 'Activities of the Radio Corporation of America' (April–July), ATT archives, Box 51.
Sarnoff, D. (1920) report to RCA president E. Nally (16 February), Museum of American History, Clark collection, Box 74.
Secane, Lt Colonel C. (1923) Signal Corps (29 June), National Archives, RG 45, Box 983.
Underdown, K. (1902) *Minutes of evidence taken before the inter-departmental committee on cable communications*, London: Wyman and Sons.
US Congress (1921) Hearings on S 4301 'A Bill to prevent the unauthorized landing of submarine cables in the United States' (hearings quoted: Frederick R. Martin, general manager of Associated Press, W. Hawkins, United Press International, and O. Young), Senate Committee on interstate commerce, 66th Congress, third session, (15 December 1920–11 January 1921).
US DOS (1922) Communiqué from the vice-consul in Marseille (6 December), DOS, M 560, Roll 141.
US Navy (1918) 'International communications to the United States', Memorandum of the Office of Naval Intelligence (14 pages), National Archives, RG 38, Box 379.
US Navy (1919) Memorandum of the Navy Communication Service (9 October), National Archives, RG 38, Box 379.
Vasseur, A. (1975) *De la T.S.F à l'électronique*, Paris: ETSF.

5

DESIGNING AND OPERATING STORM WATER DRAIN SYSTEMS

Empirical findings and conceptual developments

Konstantinos Chatzis

Introduction

This chapter offers a historical review of design and operation practices (in the broadest sense: technical and managerial know-how; organizational patterns; and relations between actors) developed in the sector of drain systems between 1850 and the present. The chapter has two main objectives. The first one is to add to the empirical knowledge of large technical systems (LTSs) by studying a system that has been little studied until now.[1] The second objective is more theoretical. The paper offers a conceptualization of the notion of 'design and operation practices', using material from the history of science and technology and from the sociology of organizations. This conceptual model was successfully applied elsewhere to the study of two other socio-technical systems, the Taylorian model of industrial organization, and the Paris subway system (Chatzis 1993a, 1995b).[2] It seems well suited to the study of other types of LTSs.

Dealing with the evolution of LTSs is not without precedent. Thomas P. Hughes in particular has already put forward a series of concepts to account for the development of such systems (Hughes 1983, 1987). Keeping Hughes's original model in mind, we intend to focus on issues specifically related to the evolution of design and operation practices in LTSs. Like Hughes's, our model of system evolution exhibits different phases: the genesis, daily routine, and crisis of a given set of practices, each characterized by specific features. As in Hughes's model, each phase involves different prominent actors: the community of engineers in the first phase; and the utility or the manufacturing firm in the second and third phases.

More specifically, we will address a number of questions about the life cycle of design and operation practices. How do the various communities of system builders manage to translate general objectives assigned to the system (e.g. protection against flooding) into actual practices (e.g. a design formula)? What are the relationships between design practices and the form of the organization that implements them? What are the connections between such practices and the various actors involved in drain systems' operation (users, normative entities such as the State, the utility company, etc.)? How does the standardization of practices lead to their crisis?

We will address these issues through the study of the three stages (genesis, routine, crisis) of the traditional design and operation practices used in drain systems. The first part of the chapter offers a factual history of these practices, whereas the second part contains the theoretical elaboration. The conclusion discusses the 'ontological' status of the notion of design and operation practices.

A history of the design and operation practices in drain systems

Producing design formulae for urban drainage systems (1850–1910)[3]

At the origin of practices: the engineer and the sewer

It was not until the middle of the nineteenth century that the problem of runoff removal became an explicit concern in European and American cities.[4] Flows of people and of goods started to increase rapidly when the population and density of urban areas increased. And so did the importance of roads, which bore these flows. Floods, which turned roads into veritable 'swamps' became less and less tolerable from a functional point of view.[5] But the reduction of permeable surfaces that resulted from growing urbanization and from the paving of streets increased the frequency and intensity of floods which also endangered lives and property. Palliative measures, such as converting the concave surfaces of road into convex ones to avoid the stagnation of rain water in the streets proved inefficient (Emmery 1834, Hederstedt 1865). What could be done?

A new actor and a technical artefact were to handle the problem: the engineer and the sewer. Underground and therefore invisible, functioning automatically through gravitation once it had been built, the sewer was able to insert itself discreetly into the social space of cities because it did not question the then predominant principle of private property (Rawlinson 1852, Bazalgette 1865, Offe and Volker 1975). Neither the installation nor the maintenance of the sewers made it necessary to move the boundary between the public and private spheres. The change would occur through the actions of urban planners when they sought to control the process of impermeabilization of land surfaces.

A discreet object, the sewer none the less caused a dual transformation. It transformed a social problem (floods and their related risks) into a technical issue (producing a technical artefact). This in turn placed engineers responsible for setting up the network in a central position and protected them against the intervention of other actors (e.g. political),[6] thus allowing them to display their specific type of rationality.

1850–1910: the antagonism between two design formulae and the long path towards standardization

Around 1850, England and France were the first countries to carry out systematic investigations into the size of drain networks. Two countries, two different national styles (Hughes 1987). In England, observation and induction held sway. The volume of rainfall in urban basins and the quantities of water removed by the existing network were repeatedly measured. Statistical treatment was used to establish average ratios between rainfalls and the corresponding flows in pipes (Porter 1986); these ratios varied as a function of the area drained, the characteristics of buildings over this area and the slope of the sewers (Rawlinson 1852, Bazalgette 1865, Metcalf and Eddy 1914, 1928). Even if a number of factors involved in the runoff phenomenon were identified (evaporation, role of the slope of the surface) accurate understanding of the relations between the rainfall and the corresponding discharge of sewers was not sought. Direct observation and a statistical treatment were considered sufficient without too much concern for causal mechanisms.

Meanwhile, on the other side of the Channel, a deductive approach was being used. Instead of accumulating observations and measurements, calculations were made to go directly from cause (rainfall) to effect (flow in pipe). The worst possible situation was imagined: a totally saturated surface after long, continuous rainfall. This type of reasoning led engineers to design the sewers so as to absorb all rainfall which might potentially fall on the area (Dupuit 1855). And later when the initial doctrine which provided for pipes that were too large was downscaled, it was still almost always through a deductive process. After having observed on one single collecting drain that the maximum flow duration was much larger than the duration of the storm which had produced the flow, Belgrand, a French engineer, introduced a new element into the analysis of the runoff phenomenon, alongside the type of soil and the slope: time. Because rainfalls have a limited duration, the furthest parts of the drained area do not contribute to maximum (peak) discharge, since the water collected there reaches the outlet only after the storm is over. Based on an assumption on mean flow velocity, Belgrand concluded that the drainage area to consider when designing a drain system should be approximately 8 km long (Belgrand 1887).

France's contribution to the science of drainage was to remain associated with the name of Belgrand for a hundred years. Belgrand's approach, which brought observations and abstract reasoning together, would triumph some thirty years later in another country: the United States. In the meantime, the English method prevailed for designing sewers. There were at first numerous measurements of rainfalls and flows of water in conduits. Then attempts were made to formalize the relationship between the two magnitudes. It was the era of 'empirical formulae'. In an imposing study including data from the largest European cities (Bürkli-Ziegler 1880), a Swiss engineer delivered to the engineering public the first of these formulae, which were ultimately generalized as:

$$Q = C\, S^{x}\, A^{y}\, I^{z} \qquad (5.1)$$

where Q represents the maximum flow to be drained, A the drained surface, S the slope of the surface, I the intensity of the largest amount of rainfall observed for the locality in question, C a coefficient depending upon the nature of the soil, and x, y, z coefficients which make it possible to adjust the formula as closely as possible to the measurements of the site (Metcalf and Eddy 1914, Chow 1963).

First developed in Europe, drainage techniques achieved their full potential in the United States. In 1880, after a stay in Europe, the American engineer Hering imported the Bürkli-Ziegler formula back to the United States in the hope of seeing it replace local practices mostly based on the English approach (Hering 1881a,b). But the first important American contribution in the field of drainage occurred in 1886, and was produced by an engineer working for the city of St Louis, called McMath. Starting from local and individual data, McMath proposed a 'method of universal validity based upon direct observation'. Convinced of the powers and virtues of induction, McMath considered that the proliferation of data that would result from the intensive application of his method would make it possible to arrive at 'a standard of capacity below which we must not go, and if the experience is extended, we [would] obtain a standard of capacity beyond which we need not go. If this could be done with data from every part of the world, the result might be a safe general rule' (McMath 1887: 180).

The year 1889 was decisive in the history of the design and operation of urban drain systems, when Kuichling, an American engineer, ushered in a new point of view on the runoff phenomenon. Kuichling developed the speculations made in 1886 by Hoxie, another engineer, who suggested that the maximal effect of torrential rainfall is attained 'when the grasp of the main channel encircles, at once, a fillet of drops from every point in the area it drains' (Hoxie 1891: 74–5). Kuichling's method differed from existing ones in several points. First, Kuichling introduced the concept of 'time of concentration', which designated the time necessary for water to flow from the most time-distant point of the drainage area to the point under observation; he was then able to organize existing data within a causal framework.

Not content merely to correlate observations between the heaviest rainfall and the corresponding runoff flow, Kuichling proposed an explanation for the origin of the maximum discharge: 'for a uniform rate of precipitation, the concentrated discharge from a given surface will become a maximum on the condition that the duration of such rate is equal to the time required for the water which falls upon the most distant point to reach the place of observation' (Kuichling 1889: 5–6).

Once the synthetic image of storm water runoff had been decomposed into its constitutive elements (maximum discharge, time of concentration, rate of precipitation),[7] it became possible to represent the runoff phenomenon by a new equation. Multiplying observations to accomplish the leap to generalization, as McMath advocated, was no longer indispensable. Armed with an understanding of the causal mechanism of the runoff phenomenon, Kuichling was able to exhaust the diversity of instances which could occur in reality, and to process the observations obtained in the light of the pre-established theoretical premises. Carefully prepared observations seemed to indicate that the maximum discharge for a uniform rainfall (whose duration was at least equal to the time of concentration) was proportional to the rainfall intensity and the amplitude of the impervious surface of the zone to be drained. In mathematical terms, this new method, immediately termed 'rational' by its proponents, took the following form:

$$Q = C\,A\,I \qquad (5.2)$$

where the variables have the same meaning as in Equation 5.1.

Let us carefully examine this formula. Aside from the original conceptual structure (the time of concentration is captured in I), it differs from 'empirical' formulae in another, very important way. While coefficient C in empirical formulae captured various features of the drainage area that could only be deduced from observation, C was equated in the rational formula to the percentage of impervious surface in the area, and was thus likely to be established *a priori* from the land-use maps. Coefficients x, y, z of empirical formulae needing on-site measurements to correlate rainfall and discharge had also disappeared.

Now, an engineer at his drawing-board only needed a zoning map and the taxonomy of land surfaces to compute coefficient C and the time of concentration corresponding to the area to be drained. Based on these two elements and pluviometric data, the engineer could calculate the discharge without actually having to measure it. Particular places were no longer the 'main variables' conditioning the validity of the mathematical formulae. Places were now characterized as a combination of a few elementary variables (time of concentration, impervious surface, rainfall intensity). The result was that the complexity of individual cases was no longer left in the hands of a local engineer whose production (a special design formula fitting local conditions) could not be supervised. The engineer could now manipulate predefined elements in a predefined way: the 'rational method' had universal validity.

The two design methods (empirical formulae vs rational method), were debated for nearly twenty years before the rational method finally became dominant around 1910.[8] Kuichling had from the beginning contested the logical form of the empirical formulae[9] and mobilized against them all kinds of arguments concerning 'engineering psychology' (e.g. the engineer was supposed to use the same empirical formula anywhere as he stood without any prior enquiries concerning its field of validity).[10] But empirical formulae held firm, mostly thanks to their flexibility which allowed them to be as effective as the rational method in any local context (Kuichling 1889, Parmley 1905, Gregory 1907). Calculations and abstractions, advocates of empirical formulae argued, were only so many deforming mirrors preventing the transparency of direct observation (Parmley 1905, Gregory 1907).

Strange as it may seem, it is a pragmatic argument which seems to have settled the question in favour of the rational method: only the rational method made it possible to design a drain system *ex nihilo*, while empirical formulae needed a pre-existing network upon which to carry out measurements. Kuichling's original formula was eventually preferred to more sophisticated versions, in particular that of Grunsky (1909). Kuichling's version was simple to use and resulted in a relative oversizing considered as a supplementary guarantee of security: it would dominate the design practices for years. Simplicity and security thus took precedence over accuracy. Regarding security and simplicity, the community of engineers soon took another step forward. While at first the rational method had not been dependent on the chosen level of protection (and acceptable risk), protection against ten-year frequency rainfall then became the standard rule for sizing networks, as supposedly the optimal choice between construction costs and damage endured.[11]

Applying the rational method: from formulae to actors (1910–70)

From a plethora of design formulae, it thus all finally boiled down, around 1910, to one standard formula, the 'rational method', set up as a standard in the field of sewer network design nearly practically world-wide – at least until the arrival of computers on the scene in the 1960s.

In the previous part we discussed formulae. Rainfall, coefficient of imperviousness and surfaces were combined to create artefacts (sewer mains, siphons, regulating devices).[12] But artefacts do not exist solely in and of themselves. They constitute active elements of organizations and human relationships (Latour 1987). Three broad categories of actors are linked around the sewerage system: citizen-consumers, utilities and normative institutions (associations of engineers, the State).[13] Let us consider them in turn.

Users' intervention is rather minimal in the field of runoff management, since users are integrated once and for all in the design through the level of protection chosen by the engineer-designer. Once the system has been designed and built, there is no interaction between the user and the system which functions automatically.

For similar reasons, utilities also play a passive role. According to the rational method, the performance of the system is primarily determined by its design. Operation is *de facto* a secondary task, aimed at preserving the initial state and capacities of the network. This dichotomy is reflected in the supremacy of normative institutions, which set the rules for the design of systems, over utilities, which merely run them. It is also reflected in the internal organization of utilities.

Because design can be carried out without any knowledge of actual operation of a given system, internal communication between utility departments (namely design, building and maintenance departments) is traditionally kept to a minimum (Chatzis and Laterrasse 1992). Moreover, the few exchanges which do occur follow well-codified channels using a formalized language (standardized technical drawings) in a sequential process: operation begins only when the design has been completed. This results in organizations in which departments are largely autonomous.

But tasks are not only isolated, they are also repetitive. A successful model of Taylorizing intellectual tasks, the rational method also sets up routines in operational procedures. Inspection visits and other maintenance procedures occur at regular intervals. This encourages the development of a culture of action characterized by a notion of order in the way the various tasks are carried out (Reid 1991, Levy 1978). To sum up, the specialization and separation of functions, together with minimum flows of information, as in Fayol's hierarchical–functional model (Fayol 1916, 1979), characterize an organization (the utility) operating on the basis of definitively established procedures, and on tasks which remain stable in the long run.

The crisis of the rational method and the quest for new design and operation practices (from 1970)

An external and internal crisis

Since the early 1970s a series of radical changes, both technological and social, have called into question the efficiency of the rational method, and have reconfigured the 'space' within which actors involved in the management of sewer systems had previously moved. Sewer systems, which purposefully provided a solution to the problem of flooding without getting caught up in land-use issues, was forced to cope with changes in urbanization. The growing impermeability of land resulting from urban spread caused a continuous increase in the volume of runoff water. Improved drainage also caused the time of concentration to reduce, and this also contributed to higher maximum discharge flow, which had resulted in more and more frequent floods since the 1960s. Changes in attitudes to the environment, and more generally increasing demands concerning the quality of urban life, made such floods unacceptable.

Faced with this situation, responses were at first in keeping with traditional practices: those in charge called for larger and larger collecting sewers intended to absorb the additional volume of water. But they came up against numerous technical and economic problems (finding room for the new sewers, facing increasing building, operation and maintenance costs in a context of low public finance).

The internal dynamics of the sewer system have aggravated the process: as systems get bigger with more, larger drains, the time of concentration of rainwaters is reduced, which tends to increase maximum discharge. The result is a spiralling process in which supply creates its own demand. A proportion of the capacities of the sewer drains are constructed just to absorb the maximum discharge created by the drains already in place: the systematic implementation of the 'rational method' is self-disrupting.[14]

A successor to the rational method: in quest of new design and operation practices

Faced with the limits of existing design and operation practices, utilities have successively tested a variety of technical solutions. New concepts for handling flows, such as the temporary storage of rainwater in retention basins, have challenged the principle of immediate runoff removal which until then had been the exclusive doctrine, and have made it possible to reduce the size of downstream sewers. Source-control techniques (porous pavement, flat roofs, infiltration tanks, swales and so on) which aim at preventing or slowing water flows into the sewers, competed directly with the notion of the network as the single technical object involved in drain systems. Computers and automated operations have made it possible to react differentially to each specific rainfall event. Based on real time measurements (and even forecasts) of rain and water flows, remote control devices (valves, siphons) are used to maximize the potential of the existing infrastructures. These technical changes have fostered major changes in the management of sewer systems.

First, the introduction of these new technologies has profoundly changed the traditional relationships between actors involved in sewer systems. For example, the new ability to react in real time to a specific rain event means that the performance of the system no longer solely depends on accurate design standards and the systematic upkeeping of civil engineering works. The performance is also based on the autonomous action of utilities. And real time control avoids standardization because it is a reaction to a unique event. Through their control over automated operations (based on new computer technology), utilities also become more active in the design process. Computer simulation, valid only for one locality (insofar as the empirical coefficients are a function of local conditions), reintroduce the local dimension (site specificity) and thus increase the autonomy of the local utility, which escapes the strict supervision of standard bodies.

In parallel, the development of source control techniques requires that the utility comes to terms with new actors influencing the performance of urban drain systems. Thus, a number of source control techniques were developed to deal with private property; their design and implementation, as well as their maintenance, fall on actors outside the utility (builders, developers, citizens). From this comes a need to develop relationships with new partners (including other utilities, such as, for example, the local highway department with regard to porous paving).

These changes also affected the context of flood risk management. Utilities in the 'rational' era were protected from social or political attacks when floods occurred because they merely implemented externally-defined standards. Since they now play a significant part in the performance of current sewer systems, utilities cannot deny reponsibility for floods any more. Moreover, elected officials face growing social pressure (angry consumers) which they tend to transfer to the utility. But the utility, deprived of the 'objectivity' of the standard rational formula, often faces difficulties in justifying its actions. In other words, it faces the problem of legitimizing its action, a problem which had been largely solved in the past by making flood risk management a purely legal matter, namely, the implementation of a standard (Chatzis 1993b). The utility is now required to meet the demands of the public and of political representatives in addition to its own operational objectives, while at the same time being able to evaluate its own performance.

Real time control has also modified the internal workings of utilities. Aside from introducing new know-how and expertise in traditional utilities (computer processors, technicians specialized in installation and upkeep of automated devices, hydraulic technicians), computerization has required increased integration within the utility. Indeed, the growing number of actors (alongside the traditional maintenance actors, new actors are now responsible for real time management) requires the development of coordination mechanisms to make individual actions compatible with each other. Since real time operation depends directly on the physical condition of the network (such as the cleaning programme, or drains technically out of order), information coming from the maintenance domain is sent (or should be sent) to the actors responsible for real time operation. Moreover, negotiation between competing technical options becomes a central element in the functioning of the utility.

The introduction of real time into the operation also modified the 'culture of action' within the utility. Aside from the technical objects, the utility must also manage rain events and all kinds of random events (breakdowns, blocked pipes, etc.) in real time. Utilities must be able to integrate the notion of time in their daily functioning, and to successfully confront the resulting constraints. This management of time necessitates a flexible organization, able to mobilize personnel rapidly and to implement actions without having to go through rigid and time-consuming administrative red-tape (Chatzis and Laterrasse 1992).

Finally, 'repetitive tasks', the major feature of the functioning of a traditional utility, are also affected by the computerization of operations. Under the requirements of real time, objectives to be reached or jobs to be done are progressively substituted for repetitive tasks, leaving a wide latitude to the actors concerning the choice of means. Obviously the less the tasks are set and their accomplishment pre-programmed, the more there will be a need for coordination inside the utility. Communication and negotiation about what used to be marginal elements have thus been transformed into a central condition for the proper functioning of sewer utilities (Chatzis 1993b).

There is a need for the utility to legitimize its action, to set up partnerships, to develop organizational patterns in line with its new role in the achievement of sewer system performance and to invent methods to foster effective and coherent actions from a large number of actors. These are the major challenges to be addressed by the new design and operation practices.

From narratives to the notion of 'design and operation practices'

Based on the historical account developed above, we are now able to elaborate the theoretical notion of 'design and operation practices'. The notion is a three-stage one, each stage corresponding to a phase in the life cycle of a given set of practices: genesis, daily routine, crisis. Let us examine each of these phases in turn.

Genesis

A set of design and operation practices is meant as a response to a social issue. Natural events (for example, floods) are given symbolic significance and become the subject of a collective discourse. Transformed into socially acknowledged problems, they demand appropriate solutions. Specific actors are then given the task of collecting the signals sent from the social arena and transforming them into a system of specific actions. In the case of storm water management, as we have seen, the community of engineers plays this role. We should thus turn to this community in order to study the genesis of a set of design and operation practices.[15]

It should be noted here that the community of engineers is not like a trade association pursuing a corporate strategy and seeking power and social legitimacy. It is rather a body aimed at centralizing and circulating expert knowledge, as well as recording and evaluating its members' activities. Communicating through professional journals and public debates, the community of engineers acts as a 'collective subject' able to provide answers to expressed needs. How do these answers take shape? Here the notion of *referential* may be usefully brought up.

The referential is the historical context within which the members of a community implement their projects; it produces, perpetuates and circumscribes what individuals are thinking and doing at one moment in history (Dreyfus 1989). But the referential does not merely express a historical force. Acting as a supplier of 'tools' for participants in a debate, it also underpins the 'communicative rationality' which allows for the formulation of solutions being sought. The notion of referential may be split into three sets of 'tools' with various degrees of systematic elaboration and codification.[16]

Most formalized among these tools are forms of 'codified knowledge'. Mathematics (e.g. statistics) and graphic symbolism (perspectives, diagrams, etc.) are the archetypes of such tools. Standardized metrological instruments ('reified theorems' in the words of Bachelard 1933: 140) and relevant theoretical elaborations (for example hydraulics in the case of drainage and sewerage) are also forms of codified knowledge. Let us note that such codified knowledge is not directly applicable to the problems for which design and operation practices are developed. Mediations which specify the conditions of application of the available 'tools' to reality are always required. This explains the discrepancies observed between the potential offered by these tools at a given time, and the extent to which they are effectively used.

The second class of tools can be called 'systems of representation'. These are precepts and general strategies for resolving problems; they promote certain methods of investigation, evaluation and testing. For example, should one observe and generalize by induction, and adopt a 'black box' attitude with respect to the phenomenon under study, or should one proceed by deduction, going from cause to effect and expliciting the cogs and wheels of the mechanisms involved? An important example of such systems of representation is the 'analytical ideal' according to which the deciphering of reality is ensured by a process of decomposition and differentiation (Cassirer 1932, 1951).

Values are the third class of elements of referentials. These relate to general attitudes regarding technical and social realities. Unlike systems of representation, which play a very active part in the shaping of practices, values act either too far upstream (as is the case with the 'ideal of automaticity'),[17] or else too far downstream when actors must choose among practices already in place. Thus, ease of implementation or of standardization can sometimes lead to choosing one set of practices, while another set is recognized as nearer to reality. Values obviously also include the moral values of equity and justice, or values with a 'humanistic' content (for example: reducing sewer cleaning, a task usually considered unpleasant), or economic values (such as fitting the solution as close to needs as possible, so as not to burden the community with excessive expense).

Institutionalization (routinization)

At given points in time during the history of a socio-technical system, the process of elaboration of practices is interrupted, giving way to a period of stability: a given set of practices becomes dominant, thus causing competing practices to fall into oblivion. This is the time for handbooks and standards.

The process of stabilization, or 'closure',[18] involves two communities: the first (the community of engineers) is characterized by a specific form of communicative rationality and by creativity; the second (the organization) follows a rationality of standards and is characterized by repetition. The organization (a utility, a manufacturing firm) implements standard practices and distributes them in the form of hierarchically organized roles; it becomes the central actor during the second phase.

Stabilization, it should be noted, does not mean that no change can occur, but changes are limited to what is required for the reproduction of standard practices. It does not mean either that the organization, although it is responsible for the implementation of standardized ('stabilized') practices, is the sole actor involved in the operation of the system: the role of other institutions, and their relations with the organization, must be carefully studied at this stage.

The organization itself may be analysed through two variables. The first is the division of labour as it is materialized in the functional structure of the organization. The second is the 'informational economy' within the organization, which in turn has two dimensions: the degree of codification of information, and its modalities of circulation (Aoki 1988, Veltz 1988).

Crisis

Crisis occurs when the current set of practices, having exhausted its possible variations, no longer manages to satisfy the aims assigned to it. Without denying the relevance of analysing crises in terms of a conflict between the system and a contingent modification of its environment, we shall none the less try to add to the concept of crisis by examining the transformation of the system under the impulse of standardized practices that remain stable throughout the second phase.

The result of a long historical process of accumulation and processing of information, standard (codified) practices paradoxically tend to overshadow their origins. Individual and collective curiosity cease when practices are standardized; standard practices simultaneously order and channel the memory of the organization, and install routine within it (Douglas 1986). The result is that standards themselves often take an active part in the crisis of a standardized set of design and operation practices. Applying a standard in a mechanical way often generates conditions which put a strain on this standard. After a period of intensive use, codified practices may no longer be adapted to the conditions which have resulted from their use. The vigilance of actors in respect of the transformations of both the socio-technical system and its environment tends to diminish under standardized practices. This process, along with the 'forgetful essence' of codified practice, explains how standards may contribute to the crisis of a standardized set of design and operation practices.

When the organization (the leading actor at this stage) becomes aware of the crisis, it tries to introduce marginal changes into the existing set of practices. However, faced with the declining performance of the old apparatus, the actors feel a strong necessity to change to another set of practices. Debate and communication come back into play. The quest for a new set of design and operation practices begins.[19]

Conclusion: design and operation practices framework – concept or reality?

We wish to conclude this empirical–theoretical investigation by addressing the following, 'ontological' question: Is the notion of design and operation practices a conceptual tool for analysing reality, or is it rather a historical reality (entity)? To what extent are these two points of view incompatible? Our answer will not be 'black and white'.

On the 'theoretical side', the notion of design and operation practices may fulfil two functions. In the first place, it may be used as an 'ideal type' (Weber 1921, 1968) for the study of the evolution of socio-technical systems. From this perspective, what is emphasized are the differences in the 'paths' of different systems under consideration and the reasons for these differences (e.g. by examining whether a particular system has undergone the three phases described above, and if it has not by examining the reasons why).

The second theoretical function ensured is to provide useful guidelines for the systematic description of the design and operation practices, while indicating the empirical points of investigation (Abrams 1982, Bonnell 1980, Skocpol 1984). We shall thus try each time to identify the actors involved in design and operation practices, to disclose their referentials, to describe the codified practices and the organizational features of the utility/manufacturing firm, and to analyse the effects of the implementation of standards on the system itself. It is evident that as an analytical tool, the concept allows neither empirical conclusions nor predictions to be drawn, as to the contents of the design and operation practices and the precise trajectory of a specific socio-technical system.

But, while acknowledging that no inherent necessity leads a socio-technical system to evolve from one phase to another, we argue that the three-phase model developed in this chapter does not result solely from an analytical need for classification and from the necessity to distinguish among different orders of reality. The model does not simply describe three different states likely to be observed in the trajectory of a socio-technical system; the three phases entertain causal relationships among themselves. Thus the second phase results from the first phase's internal dynamics of evolution. As we have seen, this first phase is criss-crossed by a communicative rationality which submits each set of practices proposed to imperatives for justification and an intense series of tests. In the same way, the crisis phase partially results from the preceding phase, given the 'forgetful essence' of standardized practices. Naturally, historical contingency plays a decisive role in the real evolution of a given socio-technical system. None the less these causal links mean that, beyond its classificatory role, the three-phase model may be considered as an empirical generalization of effective trajectories (Giddens 1984). It can thus be interpreted as the probable implementation of an oriented, non-arbitrary, albeit non-deterministic, process.

Notes

1 Summerton (1994) includes a review of LTS research.

2 See Chatzis (1993a, 1995b). Taken together, these three systems constitute an empirical field sufficiently heterogeneous that the theoretical propositions set forth may be controlled (Skocpol 1984). So as not to encumber the text of the chapter, the bibliography is a starting point only. For more details, both theoretical and factual, the reader should refer to the author's bibliographical essay (Chatzis, 1993a).

3 Let's make it clear once and for all that our historical review deals solely with the quantitative side of storm-water drainage (protection against overflowing). We do not deal with issues such as public health (and the pollution carried by sewer flows). On these subjects, see Tarr (1979) and Hamlin (1987).

4 For the period before this, see Metcalf and Eddy (1914).

5 The development of drain systems in the USA has considerably benefited from municipalities' awareness that such systems gave them a clear advantage over other municipalities in attracting new firms (Tarr 1988).

6 Engineers were responsible for the design of sewer systems. The diffusion of this technical solution was obviously in the hands of political actors.

7 On the relation between the intensity and the duration of rain, see Chatzis (1995a).

8 We shall be brief here as in the historical account of how the two approaches to storm-water design were developed, and shall thus not enter into the details of the discussion (types of arguments, rhetoric used) which was to oppose them for 20 years (see Chatzis 1993a: 187–275).

9 Kuichling commented that, according to certain empirical formulae, the ratio 'drained water over quantity of rainfall' diminished when the intensity of precipitation increased, which went against the laws of hydraulics (Kuichling 1889: 26).

10 Ibid: 27. According to the author, the rational method, on the contrary, by its analytical and decomposable structure, opened up a process of appropriation and amelioration by drainage practitioners.

11 The 'ten year' rule means that the builder of the network agrees to face the risk that all or part of the urban zones will be flooded on average once in ten years (Chatzis 1995a).

12 On the evolution and functioning of these objects see Chatzis (1993a: 262–75).

13 It should be pointed out that standardization of practices followed two main paths. The first, dominant path was used in the United States, where standardization was the work of the American Society of Civil Engineers (ASCE) which publicized it in the entire country through the publication of handbooks. The other path was used in France, where the involvement of the state administration in the standardization process was much more active: the government published and circulated a booklet on the subject in 1949 (Ministère de la Reconstruction et de l'Urbanisme 1949).

14 Also the increasing size of drained areas challenges the rational method: the fact that they were working on reduced surfaces (not more than about 200 acres) and consequently with relatively short times of concentration made the hypothesis of the homogeneity of rain over time and space plausible. It was this reasonable conjecture, along with the functional independence of each emissary (each one serving an autonomous catchment basin from a hydraulic point of view) which gave the rational method its realistic character (and its operational efficiency) (see Metcalf and Eddy 1928).

15 The central place occupied by engineers in the design of the three socio-technical systems we studied is not specific to these systems. As soon as we deal with technical elements and organizational expertise of a certain dimension, we must involve engineers.

16 Adapting to our purposes the ideas of Kuhn (1970), Laudan (1977) and Elkana (1981).

17 The ideal of automation is thus defined by Mayr: Have nature carry out, by herself, what one wants to do, with the help of an astute connecting of dependency links between the elements, both technical and human, which make up the system to be controlled, so that once the first impulse has been given by the engineer, the system will function all by itself (see Mayr 1986).

18 Tristram Engelhardt Jr and Caplan (1987) contains fruitful discussions about 'controversy and closure' in science and technology (see especially McMullin: 'Scientific controversy and its termination'). The reader will have noted that the controversy about sewer design formulae involves one single actor (the community of engineers). It is obvious that the structure of the 'controversy' is much more complex and 'closure' is correspondingly more difficult to attain there, when more than one actor is involved. On 'relevant groups' and 'stabilization and closure mechanisms' see also Bijker (1995).

19 The transition from the phase of elaboration of design and operation practices (first phase) to the phases of institutionalization and crisis (second and third

phases) marks a change in our strategy of analysis and description. In the first strategy, applied to the study of the first phase, we act as a 'virtual participant', using the same vocabulary and categories as the actors involved. In the second strategy we behave like an 'observer' elaborating a specific 'descriptive language'; the second strategy is used in the study of the second and third phases (see Habermas 1984, 1987).

References

Abrams, P. (1982) *Historical Sociology*, Ithaca (NY): Cornell University Press.

Aoki, M. (1988) *Information, Incentives and Bargaining in the Japanese Economy*, New York: Cambridge University Press.

Bachelard, G. (1933) *Les Intuitions atomistiques*, Paris: Ed. Boivin.

Bazalgette, J.W. (1865) 'On the main drainage of London and the interception of the sewage from the river Thames (with discussion)', *Proceedings of the Institution of Civil Engineers* 24: 280–358.

Belgrand, E. (1887) *Les travaux souterrains de Paris. 2ème partie: les égouts*, Paris: Dunod.

Bijker, W.E. (1995) 'Socio-historical technology studies', in S. Jasanoff, G.E Markle, J.C. Peterson, and T. Pinch (eds) *Handbook of Science and Technology Studies*, London: Sage Publications.

Bonnell, V. (1980) 'The uses of theory, concepts and comparison in historical sociology', *Comparative Studies in Society and History* 22: 158–93.

Bürkli-Ziegler, A. (1880) 'Grösste Abflussmengen bei städtischen Abzugkanälen', *Technische Mittheilungen des Schweizerischen Ingenieur & Architekten-Vereins* 14, Zürich.

Cassirer, E. (1951) (1st edn 1932) *The Philosophy of Enlightenment*, Princeton: Princeton University Press.

Chatzis, K. (1993a) *La régulation des systèmes socio-techniques sur la longue durée*, Marne-la-Vallée (France): Ecole nationale des ponts et chaussées, unpublished PhD dissertation.

—— (1993b) 'Histoire de la régulation des réseaux d'assainissement', in B. Barraqué (ed.) *La ville et le génie de l'environnement*, Paris: Presses de l'Ecole nationale des ponts et chaussées.

—— (1995a) 'La pluie (1870–1930): petite histoire d'une modélisation', *Annales des Ponts et Chaussées* 73: 58–68.

—— (1995b) 'Hommes, objets, organisations 1900–1990: un siècle de régulation dans le métropolitain parisien', *Flux, Cahiers scientifiques internationaux Réseaux et Territoires* 20: 12–26.

—— and Laterrasse, J. (1992) 'Des infrastructures normalisées à la régulation des flux', *Culture Technique* 26: 291–6.

Chow, V.T. (1963) 'Hydrologic determination of waterways areas for the design of drainage structures in a small drainage basin', *Engineering Experimental Station Bulletin* 462, University of Illinois (USA).

Douglas, M. (1986) *How Institutions Think*, New York: Syracuse University Press.

Dreyfus, H.L. (1989) 'De la mise en ordre des choses: L'Etre et le Pouvoir chez Heidegger et Foucault', in *Michel Foucault philosophe*, Paris: Seuil.

Dupuit, J. (1855) *Programme pour la rédaction du projet de distribution d'eau et d'assainissement de Paris*, Paris: Thunot.

Elkana, Y. (1981) 'A programmatic attempt at an anthropology of knowledge', in E. Mendelsohn and Y. Elkana (eds) *Sciences and Cultures*, Dordrecht: Reidel.

Emmery, H.C. (1834) 'Egouts et bornes-fontaines; entrées d'eau sous galerie; relief favorable à la circulation; lavage des ruisseaux en eaux vives', *Annales des Ponts et Chaussées* 1: 241–86.

Fayol, H. (1979) (1st edn 1916) *Administration industrielle et générale*, Paris: Dunod.

Giddens, A. (1984) *The Constitution of Society*, Cambridge (MA): Polity Press.

Gregory, E.C. (1907) 'Rainfall and run-off in storm-water sewers (with discussion)', *Transactions of the American Society of Civil Engineers* 43: 458–510.

Grunsky, C.E. (1909) 'The sewer system of San Francisco and a solution of the storm-water Flow Problem (with discussion)', *Transactions of the American Society of Civil Engineers* 45: 294–422.

Habermas, J. (1984, 1987) *The Theory of Communicative Action*, Boston (MA): Beacon Press (Vol. 1: 1984, Vol. 2: 1987, originally published in German in 1981).

Hamlin, C. (1987) *What Becomes of Pollution?*, New York and London: Garland.

Hederstedt, B. (1865) 'An account of the Drainage of Paris', *Proceedings of the Institution of Civil Engineers* 24: 257–79.

Hering, R. (1881a) 'Report of the results of an examination made in 1880 of several sewerage works in Europe', *National Board of Health Bulletin*, supplement 16, Washington (DC).

—— (1881b) 'Sewerage systems', *Transactions of the American Society of Civil Engineers* 10: 361–86.

Hoxie, R.L. (1891) 'Excessive rainfalls considered with special reference to the occurrence in populous districts (with discussion)', *Transactions of the American Society of Civil Engineers* 25: 70–118.

Hughes, T.P. (1983) *Networks of Power: Electrification in Western Society 1880–1930*, Baltimore (MD): Johns Hopkins University Press.

—— (1987) 'The evolution of large technological systems', in W.E. Bijker, T.P. Hughes and T. Pinch (eds) *The Social Construction of Technological Systems*, Cambridge (MA): MIT Press.

Kuhn, T.S. (1970) *The Structure of Scientific Revolutions*, Chicago: University of Chicago Press.

Kuichling, E. (1889) 'The relation between the rainfall and the discharge of sewers in populous districts (with discussion)', *Transactions of the American Society of Civil Engineers* 20: 1–66.

Latour, B. (1987) *Science in Action. How to Follow Scientists and Engineers Through Society*, Cambridge (MA): Harvard University Press.

Laudan, L. (1977) *Progress and its Problems: Toward a Theory of Scientific Growth*, Berkeley and Los Angeles: University of California Press.

Levy, C. (1978) *Les conditions de travail des égoutiers parisiens et la grève d'automne 1977*, Paris: CNAM.

Mayr, O. (1986) *Authority, Liberty and Automatic Machinery in Early Modern Europe*, Baltimore and London: Johns Hopkins University Press.

McMath, E. (1887) 'Determination of the size of sewers', *Transactions of the American Society of Civil Engineers* 16: 179–90.

Metcalf, L. and Eddy, H.P. (1914) (2nd edn 1928) *American Sewerage Practice*, New York: McGraw-Hill.

Ministère de la Reconstruction et de l'Urbanisme (1949) 'Instruction technique relative à l'assainissement des agglomérations', CG 1333, Paris: Imprimerie Nationale.

Offe, C. and Volker, R. (1975) 'Theses on the theory of the state', *New German Critique*: 139–47.

Parmley, W.C. (1905) 'The Walworth sewer, Cleveland, Ohio (with discussion)', *Transactions of the American Society of Civil Engineers* 55: 341–412.

Porter, M.T. (1986) *The Rise of Statistical Thinking 1820–1900*, Princeton (NJ): Princeton University Press.

Rawlinson, R. (1852) 'Drainage of towns (with discussion)', *Proceedings of the Institution of Civil Engineers* 12: 25–109.

Reid, D. (1991) *Paris Sewers and Sewermen: Realities and Representations*, Cambridge (MA) and London: Harvard University Press.

Skocpol, T. (ed.) (1984) *Theory and Method in Historical Sociology*, Cambridge: Cambridge University Press.

Summerton, J. (1994) 'Introductory essay: the systems approach to technological change', in J. Summerton (ed.) *Changing Large Technical Systems*, Boulder (CO): Westview.

Tarr, J. (1979) 'The separate vs combined sewer problem. A case study in urban technology design choice', *Journal of Urban History* 3: 308–39.

—— (1988) 'Sewerage and the development of the networked city in the United States 1850–1930', in J.A. Tarr and G. Dupuy (eds) *Technology and the Rise of the Networked City in Europe and America*, Philadelphia: Temple University Press.

Tristram Engelhardt, H., Jr and Caplan, A.L. (eds) (1987) *Scientific Controversies. Case Studies in the Resolution and Closure of Disputes in Science and Technology*, Cambridge (MA): Cambridge University Press.

Veltz, P. (1988) 'Rationalisation, organisation et modèles d'organisation dans l'industrie', in P. Cohendet, M. Hollard, T. Malsch and P. Veltz (eds) *L'après-taylorisme*, Paris: Economica.

Weber, M. (1968) (1st edn 1921) *Economy and Society: An Outline of Interpretive Sociology*, New York: Bedminster Press.

Part 2

NEW CHALLENGES

6

POWER PLAYS

The politics of interlinking systems

Jane Summerton

Introduction

The expansion of large technical systems by linkage with similar systems is fundamental to the developmental patterns of such systems and a dominant theme in the existing literature (see e.g. Hughes 1983, Caron 1988, Salsbury 1988). Historically, three types of linkages or interconnections can be discerned. The first is *territorial coupling of autonomous systems* ('meeting of systems') whereby independent regional or national systems of railway, electricity, or telephony have been physically connected and technically standardized over political borders. Although technically coupled, the respective systems are typically not organizationally integrated. Instead they retain their autonomy and individual identities.

By the second type of linkage, the physical connection of wires, grids or rails takes place within the framework of *full system integration* ('marriage of systems') that includes organizational merger. Historically, local systems have often been merged to form regional ones, which in turn have typically been consolidated into national systems (Hughes 1983, Armstrong and Nelles 1986). The result is an expanded system with a new system identity that often reflects the existing power relations and distribution of resources between the individual systems at the time of the merger (Robischon 1994). Corporate acquisitions, for example, can be organized as subsidiary companies, in which the new subsidiary must align with the parent company's identity and organization.

A third pattern in system linkage reflects an intermediate position, in an organizational sense, between territorial coupling of autonomous systems on the one hand, and full system integration, on the other hand. In this case, far-reaching corporate partnerships form the basis for *long-term alliances and linkages between systems* that are characterized by shared technical artefacts and organizational resources (including mechanisms for joint decision-making) while still retaining the essential identity and autonomy of the individual systems.

For example, elements of this type of alliance can be found in railways (e.g. 'confederations' of interconnecting railroad systems in the late 1800s) as well as in electricity (e.g. regional power pools and 'planned grids' that linked the resources, loads and investments of cooperating local systems in the 1920s, see Hughes 1983: 324–34). Although not as common as mergers, electricity alliances such as the Pennsylvania–New Jersey Interconnection integrated the artefacts of the allied systems and had joint financing of projects, while each of the systems continued to be organizationally and financially separate. As Hughes notes, such alliances 'brought the *economic benefits* of a large system and at the same time preserved the utilities' corporate *identities*' (Hughes 1983: 332, emphasis added).

In the 1990s, corporate managers in electricity are again shaping trans-border alliances that are based on tight coupling of technological artefacts, while at the same time allowing each of the participants to retain extensive control, autonomy and their essential system identities. What motives underlie the emergence and shaping of these alliances? What characterizes their design and internal dynamics? And what can they tell us about the politics of interconnection in sociotechnical systems and networks?

This paper will analyse processes, motives and designs in emergent alliances in large technical systems through two case studies of contemporary alliances in Nordic electricity. One is a transnational alliance, while the other is a regional linkage among previously local systems.

The paper is organized as follows. In the next section, sociotechnical alliances and networks – their motives and dynamics – are briefly explored using concepts from selected areas of social science theory. Thereafter the current reconfiguring of electricity systems will be described as backdrop to the two case studies, which are analysed in the following two sections. The paper concludes with a discussion of what these alliances indicate about 'economics', 'politics' and power in the interlinkage of systems and networks.

Networks and alliances: forms, motives, dynamics

Mergers, coalitions, and alliances – as different forms of networks – differ in their dynamics and properties. In a merger, two or more organizations consolidate their operations, resulting in the emergence of one expanded system with a 'new' identity. When one organization clearly dominates in power and size, this identity will be likely to express the practices, culture and institutionalized rule systems of the dominant actor. These dynamics are somewhat different from those of alliances. As Williamson notes:

> The alliance is a hybrid form of organization that is located between coalitions and mergers in its economic properties and differs from both in political respects. As compared with coalitions, alliances are more formal, have longer projected durations, and are normally

> thought of as taking place between rather than within firms. . . . As compared with mergers, *alliances preserve a greater degree of autonomy* among the membership.
>
> (Williamson 1990: 7, emphasis added)

Alliances enable corporate actors to retain autonomy while also gaining the benefits of collaboration such as shared access to resources and spreading of risks.

Risk spreading has been a salient motive behind joint ventures, a particular form of alliance[1] which has been termed 'the shape of post-war systems technology' (Bugos 1996: 18). As Bugos points out, joint ventures are shaped 'more by fear than greed', i.e. they are often driven more by perceived risks in developing new technologies or entering new markets than by profit motives. When faced with market uncertainty caused by, for example rapid internationalization, alliances such as joint ventures can be a way to gain flexibility and needed learning time. They are often used specifically in high capital-intensive technical systems as a means of sharing complementary technical skills and handling liability issues (Bugos 1996: 5).

The theme of complementarity as an incentive to building alliances is also a prominent theme in the literature on large technical systems. In analysing motives and principles behind the expansion and interconnection of electricity systems, Hughes (1983) shows how diversity and complementarity guided the direction and design of system growth. In designing their expanding networks, managers sought to achieve an ideal mix of artefacts and resources by interconnecting power plants whose energy sources could be used complementarily (Hughes 1983: 367). An ideal mix might include, for example, hydropower plants, urban biomass-fuelled cogeneration plants, and conventional coal-fired power plants. The result of this diversity, or 'economic mix', was enhanced economy and reliability. Similarly, managers sought systems with a different load profile ('load factors') than their own, matching peaks and valleys in consumption as a means of utilizing installed capacity as economically as possible. Thus Hughes points to the importance of *technical–economic complementarity and economic efficiency* as basic motives behind the interlinking of sociotechnical systems and networks.

Complementing in some ways the systems approach, actor-network theory (see e.g. Callon 1986a, Latour 1987, Latour 1988, Law 1987) emphasizes that sociotechnical networks are to be understood as Machiavellian instruments of power. Goal-directed entrepreneurs or 'heterogeneous engineers' (Law 1987) continuously strive to enrol and control other actors as allies to suit their own purposes. The entrepreneurs of actor-networks work on many fronts simultaneously, continually striving to extend and defend their control. If successful, the entrepreneur ultimately shapes an all-encompassing actor-world, defined as the 'world of entitities' created by the operation of an actor-network (Callon 1986a). Thus in actor-network theory, sociotechnical alliances and networks are the expressions of multifaceted strategies by which entrepreneurs of technology seek to *expand their power and increase their control* over other actors, entitites and resources that are crucial to reaching their own goals.

Actor-network theory is unequivocal about the structure and internal dynamics that characterize the design and operation of such networks. Rather than reciprocity among actors, an actor network is characterized by one-sided politics of domination. The enrolled actors sacrifice autonomy and control, are given a new identity that is imposed and stabilized by the entrepreneur (Callon 1986b), and become subservient – if inherently unreliable – entities (Callon 1986b, Singleton and Michael 1993). In actor-networks, control by the dominant actor is accompanied by the loss of autonomy by all others.

Context of change: electricity systems in a climate of uncertainty

Before exploring the case studies of contemporary alliances in electricity, we need to understand the context within which these alliances are being developed. Until recently, the core institutional feature of electricity systems – like many other large technical systems – was tightly-coupled, hierarchial and monopoly organization (Hughes 1983, Armstrong and Nelles 1986). Electricity systems in most western countries were distinctly national in scope and identity. Power relations among actors with different roles within the system (e.g. producers, distributors, consumers, regulators) were, at least generally speaking, stable as a result of decades of institutionalization and formally mandated, routinized practice.

This stability is currently being disrupted in electricity systems in many areas, notably Great Britain, Norway, Finland, Sweden, and parts of the United States (Hirsch 1989, Summerton and Bradshaw 1991, Midttun 1997). Traditional monopolies are being challenged by public policies aimed at reformulating the institutional 'rules of the game' which have guided transactions among system actors for many years. In some areas, national boundaries are being dissolved as electrons, flashed through multimillion dollar underwater cables, flow across newly-opened political borders (e.g. to eastern Europe, within the European Union), thereby changing nationality, price and cultural patterns of use.

The Swedish electricity system is currently being reshaped by two forces. The first is *internationalization*. Like most electricity systems, Swedish electricity has until recently been distinctly national in scope, excepting the long-standing Nordic power exchange through Nordel, the association for Nordic electric cooperation (Kaijser 1991). In recent years, however, utilities have increasingly expanded over national borders, stimulated in part by overcapacity and the saturation of domestic markets, and in part by political processes. These processes include the restructuring of energy markets in Norway, Finland, and Sweden and the latter's entry into the European Union.

In addition, in early 1996 a common Swedish-Norwegian 'electricity exchange' was established as a forum for the buying and selling of electricity. To facilitate export of power to continental Europe, numerous new high-capacity transborder cables have been installed or expanded.

Equally significant, a number of large utilities have established partnerships (joint ventures, cross-ownerships, purchases of stock) with foreign utilities. One of the earliest was Electricité de France (EdF) which in 1994 purchased company stock in Sweden's second largest company Sydkraft, forming an organizational linkage across national borders without either territorial integration or technical coupling of artefacts or loads. Since then, several foreign utilities have acquired partial or total ownership in Swedish utilities which in turn have also acquired foreign companies. Internationalization has dissolved national system borders.

The second factor that is reshaping Sweden's electricity system is *institutional restructuring* of the domestic market. In January 1997, reform legislation was enacted to deregulate (or more accurately, reregulate) the electricity market. The guiding principle was to establish a competitive market for generation and sales, while continuing to govern distribution and transmission as regulated monopolies. The reform means that all owners of regional grids and local distribution networks are required to grant open access to their grids/networks on reasonable and non-discriminatory terms, subject to regulatory oversight.[2] Furthermore, generation and sales must be legally separated from distribution: all energy companies are legally required to separate generation and sales of electricity, on the one hand, from electricity distribution, on the other.

In addition to the competitive scramble for foreign markets and the flux of reregulation, Swedish electricity managers in the 1990s must also deal with a third source of turbulence, namely *political uncertainty about the future of nuclear power*. Currently providing about 50 per cent of total power but deteriorating with age, the country's 12 nuclear power plants are continual topics of ongoing political controversy and public debate. In the complicated drama of high-level political posturing, it is far from clear 'if, when and how' Sweden will ultimately carry out its 1980 referendum decision to phase out all nuclear power by 2010. With the long-term viability of the resource base still challenged, electricity managers need to secure their access to reliable sources of power (notably cogeneration) which can complement existing hydro resources in a diverse production mix.

In summary, the three factors just noted – internationalization, institutional restructuring, and controversy about nuclear power – contribute to a climate of considerable uncertainty for utility managers. How will the electricity system of the future be configured, what actors will constitute its power structure and what measures can be taken to deal with these changes?

The reconfiguring of institutional and territorial boundaries challenges pre-existing power relations among system actors and their traditional mechanisms of control (e.g. monopoly concessions, proprietary grid access, and centralized ownership of core artefacts). When formal mechanisms of system control are altered or dismantled, system builders can be expected to seek alternative means for ensuring and extending their power. As just noted, one such means which is particularly prominent in European electricity in the 1990s is the shaping of strategic networks and alliances across borders and among new constellations of actors.

The following sections focus on case studies of two such strategic alliances, namely a *transnational linkage* between the largest electric utilities in Sweden and Denmark (Vattenfall and Sjaellandske Kraftvaerker, respectively) and a *regional linkage* between local energy companies in Sweden (the municipalities of Linköping and Mjölby).

Case 1: Transnational alliance between Swedish and Danish utilities

The cross-ownership alliance between the largest electricity companies in Sweden and Denmark – namely Vattenfall and Sjaellandske Kraftvaerker (SK Power) respectively – links three national electricity systems with complementary artefacts and resources. These are: (1) the Swedish hydro and nuclear system; (2) the Danish cogeneration and renewable energy system; and (3) the German conventional power/coal systems. Proudly proclaimed by its participants as a 'new chapter . . . in Nordic energy history' (*Annual Report*, SK Power 1994: 15), the alliance includes four main activities: the joint ownership of power plants; exchanges of power; shared access to transmission capacity; and cooperation in research and development.

Background: 'Putting technical differences to good use'

On a corporate level, Vattenfall and SK Power have known each other for a long time. Among other things, the two utilities have a long history of cooperation through Nordel, in which both have participated since its establishment in 1963. The new commercial alliance is a meeting of national giants: Vattenfall delivers half of Sweden's total electricity demands (about 74 TWh) and SK Power dominates Eastern Denmark with 80 per cent of total installed capacity and annual deliveries of about 12 TWh.

Although they are both public utilities, Vattenfall and SK Power are characterized by significant differences in size, organizational format and scope of operations. Owned solely by the Swedish state, Vattenfall has been a limited liability company since 1992, when its former status as a state power board was changed. As parent company to over eighty wholly and jointly owned subsidiaries, Vattenfall has a whopping operating revenue of over

24 billion SEK (about US$ 3.4 billion) and some 9,000 employees (1994). SK Power is considerably smaller, with an operating revenue of just under 4 billion DEK in 1994 (US$ 7,500,000) and 1,400 employees (*Annual Report*, SK Power 1994: 33). At the time of writing this, the utility is owned by three mainly municipal energy distributors and one municipality in Eastern Denmark, which are organized as a uniquely Danish entity known as an 'intressesällskap'. In addition, SK Power owns 80 per cent of total shares in Elkraft, the energy cooperative which coordinates all fuel supply in Eastern Denmark.

Most important, the two corporations are highly different in the artefacts and physical resources which sustain their systems. Vattenfall's system is based almost entirely on hydropower stations (9,100 MW of installed capacity) and large nuclear power stations (5,800 MW). The system at present does not have, however, extensive plants for combined heat and power generation, CHP. In contrast, CHP plants are the backbone of SK Power's system, which consists of four large power stations (primarily coal-fired, heavy fuel oil, some natural gas) and seven small-scale CHP plants (gas, straw, other). The Danish system in turn lacks, however, hydro power. The expressed purpose of the alliance is to utilize these technical differences in strategic and economically profitable ways.

Smooth path to partnership?

Although it is difficult to gain access to detailed sources of information about the content and character of the negotiations between Vattenfall and SK Power, basic contours in the process can be discerned. Executives of the two companies reportedly discussed the idea informally when attending a Nordel meeting in early 1992. Strategic areas of mutual interest were subsequently identified and reportedly confirmed at a meeting of leading executives from each company in November 1992 (*Nyhedsmagasinet SK* 1993: 4–5).

Negotiations were kept strictly secret due to their commercial character and to the fact that one of SK Power's owner companies is on the stock market, which meant that any news leaks could influence stock ratings. Only a few key executives and managers appear to have been involved in the process, which reportedly proceeded on the basis of perceived trust and personal compatibility among the corporate leaders involved. In August 1993, executives of the two companies signed a formal letter of intent and made a public announcement about the cooperation. Eight months later, the chief executives signed the full contract at Vattenfall's headquarters in Stockholm.

The content of the alliance can be summarized in four points (*Nyhedsmagasinet SK* 1994: 4–5):

- a joint company has been formed for building and operating an advanced, multi-fuel combined heat and power plant outside of Copenhagen;

- a joint company has been formed for owning and operating certain hydropower stations along the large Indal River in northern Sweden;
- Vattenfall has acquired access to one-third of total transmission capacity ('transit rights') on a 600 MW transborder transmission cable (KONTEK) that is currently being constructed between eastern Denmark and northern Germany; and
- Vattenfall and SK Power have committed themselves to cooperation on large research and development (R&D) projects.

The formal organization of the alliance appears in Figure 6.1.

It is important to note that the above areas are the 'visible' or outward expressions of the alliance, as reflected in the official rhetoric of corporate materials (press releases, brochures, company newsletters, certain documents). It is not unlikely that the alliance is also motivated by strategic exchanges and agreements in substantive areas which are not made available to public scrutiny.

Motives: technology, economics and politics?

Vattenfall and SK Power emphasize that their alliance is built on a shared vision of utilizing their systems' technical diversity and complementarity as a

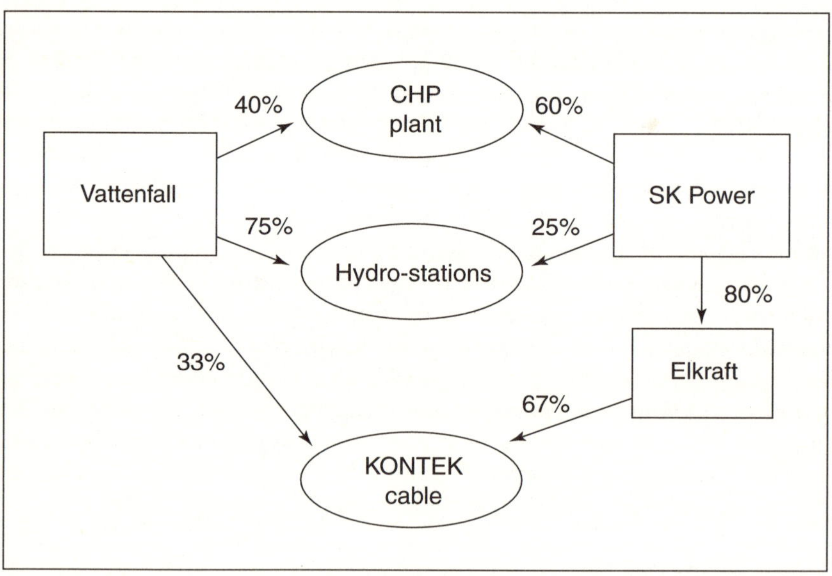

Figure 6.1 Schematic organization of the Vattenfall–SK power agreement
Note Percentages indicate shares of stock in the joint companies that own and operate the respective entities.

means of maximizing economic efficiency and increasing the competitiveness of the respective systems. This motive of utilizing technical complementarity for economic gain is expressed in several ways by corporate executives.

First, the complementarity in *technical artefacts*, i.e. Vattenfall's hydro power and SK Power's CHP, forms the expressed basis for the coooperation. For Vattenfall, access to CHP is important as a source of base load power, thus increasing system reliability in dry years or periods of nuclear problems (Elkraft Annual Report 1994: 6). For SK Power, hydro is valuable not only as peak power but also as reliable, low cost power that can be expected to be viewed as environmentally acceptable even on future markets (*Nyheds-magasinet SK* 1994: 5). Thus an important expressed motive behind the cooperation, in keeping with Hughes' (1983) findings, is to reap the benefits of an improved *economic mix*: by interconnecting their diverse power plants, managers are able to exchange power as needed and thus use these plants in ways that make most economic sense.

Second, the complementarity of the two systems' *geographic positions* and their relation to load is referred to as a strategic asset in maximizing economic returns. The managing director of SK Power appears to be well aware of his company's favourable geographic location:

> We are sitting right in the middle of things. Even though Denmark is a small country in Europe, and Sjaellandske Kraftvaerker only one-sixth or one-seventh the size of Vattenfall, we can certainly benefit from our commanding geographical position in Europe.
>
> (Cited in *Euroenergy* 1994a: 49)

SK Power's geographic position is important because its system links the Scandinavian systems with those of the rest of continental Europe. Access to the SK Power system thus gives Vattenfall closer proximity to lucrative continental markets, and new cables give guaranteed access to transborder transmission capacity to reach these markets. Electricity trading on foreign markets is important because it allows utility managers to take advantage of variations in the load profiles of Nordic and continental systems. Even differences of one hour peaks in consumption are economically significant, pointing to the advantages of *diversity in load* ('load factor', see Hughes 1983) as a motive for interconnecting systems with different geographically-contingent profiles.

A third expressed motive of utilizing technical complementarity in economically beneficial ways concerns the two companies' respective *areas of engineering competence*. Vattenfall engineers and managers essentially lack knowledge and experience of CHP, which is a core area of professional competence for SK Power. Participation in SK Power's new advanced CHP plant project is thus expected to strengthen Vattenfall engineers' engineering and managerial competence in CHP construction and operation, which in turn can be a future strategic asset. 'This can give us invaluable experience for when we have to build our own plants', comments the head of Vattenfall's electricity supply division, Anders Hedenstedt (cited in *Euroenergy* 1994b: 43).

Despite the emphasis on the economic rationale for the alliances, there are, however, clearly other motives as well. The expansion and interlinkage of systems across national borders is an integral part of managers' long-term scenarios for expanding their systems. Says Lennart Lundberg, vice president of Vattenfall: 'What remains, as far as system building is concerned, is to connect the Nordic systems with the Continent and to link together Eastern and Western Europe' (cited in *Euroenergy* 1994c: 18).

In the terminology of actor-network theory, goal-directed managers are thus striving to shape all encompassing 'actor-worlds' that link not only Nordic electricity systems but also those of east and west Europe. Foreign systems and markets are to be connected as part of carefully planned Machiavellian strategies for expansion:

> We can't expand more in Sweden now, all possibilities are exhausted on the production side . . . Foreign markets provide a possibility for expansion, while at the same time enabling us to vitalize the company, maintain our technical competence in production, and spread risks.[3]

For managers, the fate of one's own system – its vitality, technical competence and risk-spreading capability – is thereby explicitly tied to the ability to expand to new international markets. Transatlantic alliances that link artefacts, loads, markets and areas of competence are viewed as a strategy for achieving this global scenario:

> If you ask colleagues in the electricity industry in other parts of Europe, they say they do not believe in these ideas (of liberation and free transmission rights of European electricity markets). And yet at the same time they are all feverishly preparing for them. That is why it is extremely important for electricity producers to make alliances and set up transmission connections now. These two factors are closely linked, of course, and cannot exist in isolation.
>
> (Preben Schou, cited in *Euroenergy* 1994a: 49)

In order for such alliances to be effective, they must be strong and contractually locked in place, so as to form a united front when dealing with competitors:

> I . . . believe in strong bilateral agreements of the type we have with Vattenfall. And it is important for the Nordic countries to stick together, even if it is on commercial terms. Entering into bilateral

> agreements will also mean that the Nordic area is large enough for us to hold our own in Continental Europe.
>
> (Preben Schou, cited in *Euroenergy* 1994a: 49)

Alliances are thus essential parts of expansionist strategies on transnational markets.

Configuration of alliance, internal dynamics

In its design, the Vattenfall–SK Power alliance is organized on the principle of joint ownership of separate companies for specific projects rather than merger or cross ownership between the two parent corporations. Why was this design chosen?

One explanation lies in political restrictions. Both VAT and SK Power are public corporations, and neither were (or are) available for sale by their respective owners, i.e. the Swedish state and the Danish municipalities. An arrangement based on cross ownership of parent companies across national borders was simply not politically viable.[4]

Also, despite all the rhetoric of reciprocity and interdependence in sharing resources, it is clear that the representatives from Vattenfall are aware that they are striking a potentially sensitive cultural and political chord with regard to control of Swedish *hydro power*. The first Vattenfall press release repeatedly underscores that SK Power will be a 'minority' holder in the joint hydro company, that this company will be a 'Vattenfall subsidiary', that Vattenfall will be responsible for maintenance, and that Vattenfall's personnel will operate the hydro-stations. This wording reflects the acute awareness that in Sweden at least, hydro is regarded as a distinctly national resource which is not readily relinquished. Any suggestion of efforts to sell the national 'heritage'[5] – and the enormous sunk costs it represents – would predictably lead to extensive public debate.

Managers' motives for joint ownership rather than merger also stem from perceived needs to retain autonomy and flexibility. This autonomy includes the freedom to shape alliances with other prospective entities in other parts of their actor-worlds, as implied by Vattenfall's chief executive officer Carl-Erik Nyquist: 'The power industry will tend to work more on the basis of alliances. In this context, we believe more in the idea of project alliances with different partners in different places than in formal connections' (cited in *Euroenergy* 1994c: 19). Shaping multiple networks and linking them to fit the entrepreneur's plans is apparently viewed as the system-building strategy of the future.

Notably, the internal structure of the alliance indicates a careful attention to formal mechanisms to achieve an acceptable balance of power between actors. Each corporation retains majority stock in its core resource while gaining significant minority stock in the core resource of its partner. The hydropower and CHP which are exchanged are essentially equivalent in capacity, and they are also designed to reflect 'national equivalencies'.[6]

Corporate executives are keen to emphasize the reciprocal nature of the alliance. A representative from Vattenfall notes, for example, that an ownership format by which SK Power would own a small percentage of Vattenfall would not have reflected the explicitly 'cooperative' nature of the alliance.

Case 2: Regional alliance between two municipal energy companies in Sweden

Until recently, municipal energy companies in Sweden have essentially not cooperated with each other in planning or managing their systems. Since the early 1990s, however, many municipal alliances have emerged in response to the changing domestic market. One of the most innovative examples is the alliance between the energy companies in the towns of Linköping and Mjölby (both in southern Sweden).

Two municipal energy companies

The municipalities of Linköping and Mjölby are located 200 and 250 km, respectively, south of Stockholm. Similarly to the Vattenfall–SK Power alliance, their energy companies reflect considerable differences in size, technical artefacts and organizational resources.

Linköping's energy company, TVL, is a large 'public works' utility that provides electricity, district heating, water, waste disposal and other services to 120,000 municipal inhabitants. The company, which is owned solely by the municipality it serves, is one of Sweden's largest producer–distributors of electricity. Notably, it has extensive resources for self-generation of power, providing 65–75 per cent of its own power needs and delivering about 1 TWh annually. The company utilizes a highly diversified resource base that includes cogeneration through the use of wood chips, peat, trash burning, coal and oil, as well as hydropower plants and experimental wind power.

Mjölby's energy company, MSE, is considerably smaller than its neighbour, supplying about one-fourth as much electricity to a much smaller town of 26,000 inhabitants. MSE's gross annual income is only one-sixth to one-seventh that of TVL, paralleling SK Power's size in relation to Vattenfall. The company lacks self-generating power plants with the exception of hydropower stations that provide less than 20 per cent of its total power needs. All other power is purchased from Vattenfall. Due to unique historical circumstances, MSE has mixed ownership that consists mainly of municipal owners (80 per cent Mjölby municipality, 11 per cent neighbouring Motala municipality), but also private shareholders (about 9 per cent, see Summerton 1992).

The two energy companies have a long tradition of cooperation, among other things through joint operation of their respective hydro-stations along the river that runs through both companies' service areas. TVL, MSE and other neighbouring companies also jointly own a (currently inactive) company for purchase of natural gas, once planned to be introduced in the region. In addition, MSE purchases all its oil from TVL and delivers all its household waste to TVL's waste burning plant.

Background: alliances for municipal self-reliance

Since the 1950s, a cornerstone of TVL's managerial policy has been to seek means of expanding the company's resources for self-power production in order to minimize dependency on large power producers while strengthening municipal self-reliance.[7] A key strategy in these efforts has been to initiate various kinds of alliances with other municipal companies which have similar visions. According to TVL's former chief executive, numerous such attempts have, however, been blocked by countermeasures on the part of the country's large producer–distributor utilities.[8]

In the early 1990s, with the political debate on deregulation of Swedish electricity gaining in intensity, new threats emerged. Many municipal energy companies were being acquired by large producer–distributors who, seeking access to local distribution networks and the potential for cogeneration offered by municipal district heating plants, paid attractive prices for assuming ownership and control of municipal systems.[9]

In some cases, the municipal owners of energy companies were eager sellers. Many Swedish municipalities were in the midst of severe financial crisis. For hard-pressed municipal *politicians* faced with dwindling state and community funds, lucrative local energy companies often represented a 'hidden reserve' of capital that could be made available by selling off the company. For energy *managers*, on the other hand, the threat of 'being sold' to a corporate utility was often not an attractive option. Instead, one alternative strategy was to ally with a neighbouring energy company. In the early 1990s, many intermunicipal mergers, consortiums and alliances began to emerge in various parts of Sweden.

One such merger was proposed in the Linköping–Mjölby area in the late 1980s. After extensive negotiations, the merger proposal failed, largely as a result of strong opposition from TVL's then chief executive, who feared that the proposed regional merger would lead to the loss of local autonomy over the system.[10] Rather than a merger, TVL's executives and managers instead had an explicit vision of shaping a regional network based on strategic linkages with other companies. The resulting network and how and why it was shaped is described in the following sections.

Regional network evolves

The alliance between TVL and MSE is embedded in a regional network with multiple linkages among actors. These actors are tied by various kinds of joint ownership. This complex network evolved in a series of separate but essentially parallel negotiation processes, each centring around a specific project or linkage.

The first linkage is based on ownership ties: since 1993, MSE is jointly owned by several municipalities in the region, including Linköping through TVL. This arrangement evolved as an explicit alternative to the merger plans noted above and was part of a scenario developed by TVL's chief executive.[11] It accommodated Mjölby politicians' interests in gaining much-needed revenue while satisfying managers' interests in retaining local autonomy and control of their systems.

Initially, the TVL chief excecutive's scenario for a new inter-municipal network did not meet with enthusiasm. In the spring of 1992, however, a rumour that Vattenfall was potentially interested in acquiring MSE set off a rapid chain of events. By the end of 1992, a concrete proposal was presented whereby Mjölby municipality would retain 55 per cent of shares in its energy company,[12] TVL in Linköping would purchase 25 per cent and a third municipal company, Norrköping's energy company, would purchase 10 per cent.[13] This design was formally approved in the spring of 1993 and went into effect soon after. Notably, an option whereby MSE would own reciprocal shares in the other companies was never discussed; these companies were so much larger than MSE that a corresponding share of their stock, it was felt, would not provide a channel of real influence for MSE.[14] Reciprocity could be better secured, it was argued, in other forms.

The second linkage in the regional power network centres around a joint TVL–MSE *subsidiary for making power purchases*, cooperating in dispatching power, and initiating other projects for enhancing the economic efficiency of the respective systems. The idea of forming a jointly-owned subsidiary was proposed by TVL's chief executive in late 1991. By combining purchasing power together, the two companies could strengthen their negotiating position *vis-à-vis* large power producers and hopefully reduce the price of power. Also, the companies could share one capacity tariff rather than paying two separate ones. Perhaps most important, managers saw the possibility of utilizing system complementarities in load and capacity needs (further described below). The executives of the two energy companies had little difficulty in reaching agreement, and the Linköping–Mjölby Power Purchase Company, LMK, was formed as a 50/50 per cent joint subsidiary in the spring of 1992.[15]

Finally, the third linkage in the regional network consists of a consortium of energy companies for *joint ownership of a large, conventional power plant*. Ownership is now shared among five different companies (see Figure 6.2). Meanwhile, the linkages between TVL and MSE also include continued cooperation in *joint management of their respective hydro-stations*.

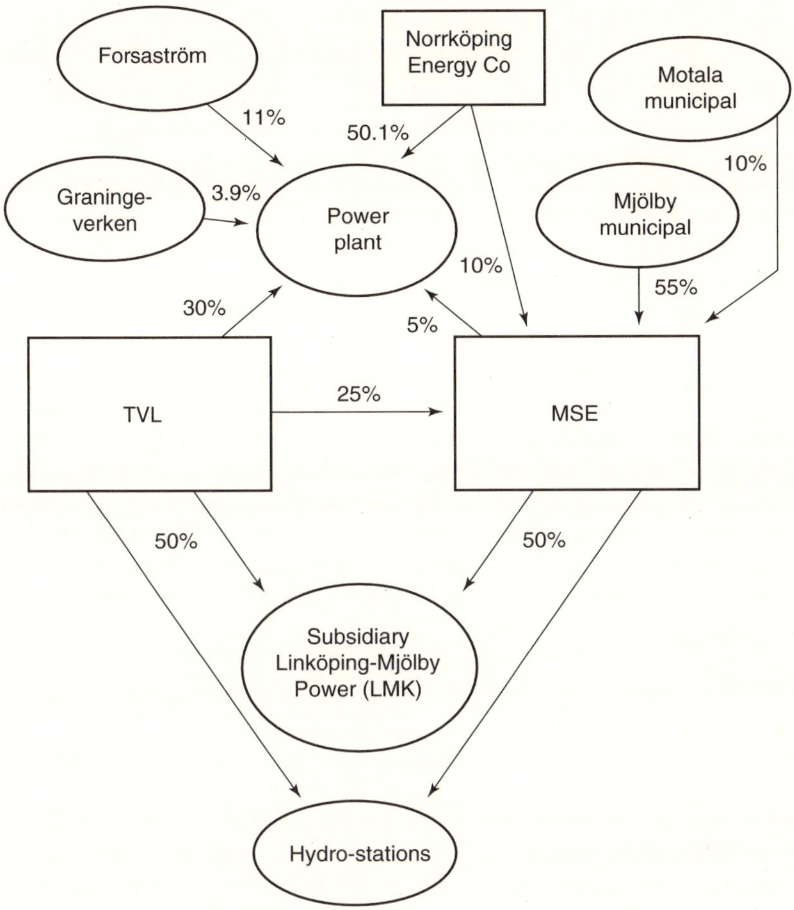

Figure 6.2 Regional network of ownership ties between TVL, MSE and other energy companies

Note Percentages indicate shares of stock in the respective companies.

The TVL–MSE alliance and the regional network in which it is embedded is schematically illustrated in Figure 6.2.[16]

Similar to the Vattenfall–SK Power alliance, the internal structure reflects an attention to formal mechanisms to achieve a measure of equality and reciprocity among two inherently unequal partners, as indicated in Figure 6.2. The joint power purchase subsidiary is owned on a 50/50 basis, and economic benefits are shared equally among the two participating energy companies. Managers also claim that the companies have equal influence over both power purchase/dispatch and hydro regulation, which are reportedly carried out on an informal, non-hierarchial 'lift the phone' basis. Mutual trust is a prerequisite for the alliance.

Motives?

What motives did the executives in Linköping's TVL and Mjölby's MSE have in forming their alliance, and what factors contributed to the way in which the alliance was configured?

Similar to the alliance between Vattenfall and SK power, the expressed motive was first and foremost to *utilize system complementarities as a means of increasing economic efficiency*. This rationale is perhaps most clearly expressed in the joint subsidiary for power purchases/joint dispatch, which entails integration of the technical resources and load profiles of the two systems. Managers pool all available power from MSE plants, TVL plants and external sources (such as purchases from Vattenfall) and apply the principle of *economic dispatch* in when and how this power is used, i.e. the lowest-priced available source of power is used first in a given situation. The joint use of resources is enhanced by system complementarities in other ways as well. When TVL has temporary overcapacity in its own power plants, for example, MSE can use this capacity as a substitute for external purchases of peak power. The two companies also have complementary peaks for electricity use. Again the principle of technical–economic complementarity is consistent with Hughes' (1983) findings.

There are also other expressed motives for the alliance. In a climate of uncertainty in which many municipal energy companies were being acquired by large power producers, the alliance evolved as an explicit *political counter-strategy* in response to clearly perceived threats. By buying company shares in MSE, it was reasoned, TVL would effectively block Vattenfall from doing so, based on the assumption that Vattenfall would not be interested in purchasing minority shares. TVL executives wanted to prevent Vattenfall from making corporate acquisitions in 'their own backyard', as one executive expressed it. Among other things, TVL's managers foresaw difficulties in attempting to operate hydro-plants jointly with Vattenfall – their 'counter-part' in power purchase negotiations – because these plants are designed to provide peak power specifically as an alternative to external purchases from Vattenfall.[17] Both the intermunicipal ownership of MSE and the power plant consortium are seen as means to strengthen municipal ownership and reduce dependency on large power producers.

The emergent regional network is thus an attempt to strengthen the participating actors' positions on a changing electricity market and to increase their power *vis-à-vis* other competitive actors. In linking heterogeneous actors and entities (e.g. power plants, politicians, energy company stock, power purchase contracts) and locking them in place by ownership ties, the managers are shaping powerful actor-networks to meet their goals. These goals include securing access to strategic resources (e.g self-generating power, distribution networks, strong negotiating positions) that are crucial to defending and expanding their operations. Through multiple alliances, each of the interlinked entities lends its force to the entire network, reinforcing its strength.

Conclusion

How can we interpret the emergent transnational and regional alliances in electricity as discussed in this chapter – their design, the motives behind them, and their implications?

As emergent sociotechnical entities, these alliances reflect sophisticated forms of system interconnection which differ in important ways from the dominant historical patterns of territorial coupling and corporate merger. In the electricity systems studied here, artefacts and organizations are linked in ways which allow actors to integrate and coordinate complementary resources and skills without sacrificing autonomy or managerial control. As we have seen, autonomy also provides managers with the flexibility to shape multiple networks.

Managers in both the transnational alliance (Vattenfall and SK Power) and the regional alliance (TVL and MSE) emphasize economic motives, reinforcing Hughes' findings concerning the importance of *technical–economic system complementarity and diversity* as guiding principles for linking systems. By exploiting complementarities in artefacts and resources, the actors argue, engineers and managers can enhance the economic efficiency of their respective systems, thereby increasing their competitiveness and maximizing profit.

The alliances reveal, however, that the 'economics' of technology cannot be interpreted apart from their inherent politics and the role of rhetoric in articulating, sustaining and expanding these politics. Like true Machiavellian princes, managers are indeed shaping powerful actor-worlds as political instruments for expanding their systems across traditional borders, increasing their power in relation to other competitive actors and extending their control in a period of uncertainty. Far more than 'economic' strategies for maximizing efficiency, the new alliances are efforts to increase managerial power to fulfil long-term scenarios.

The configuration of the alliances emphasizes interdependence, trust and reciprocity among actors, as indicated in both managerial rhetoric and formal mechanisms to insure a measure of symmetry in the distribution of power among actors. If the ideals expressed in the formal design actually reflect the new relations among actors, this would indicate – in contrast to the politics of domination emphasized by actor-network theory – that *the weaker actors of the alliance are empowered and the stronger actors are not privileged*. Reciprocity and trust would replace subservience, shared power would replace unilateral control.

Both the transnational alliance and the regional alliance must, however, be understood within broader agendas of power. They provide a means for corporate managers to *handle political restrictions* on public companies (e.g. restrictions on selling company shares) and *political pressures* on such companies (e.g. pressures to sell companies to generate municipal revenue). Transborder alliances also *challenge the jurisdiction of traditional regulatory institutions*. For example, it is not clear that a power plant which is jointly owned by utilities in two different countries (or two different municipalities) can be readily regulated by the respective national authorities (or municipal authorities) – despite 'public' ownership (McGowan, this volume). This aspect could have significant implications in the case of environmentally contested nuclear power or hydro-plants. In addition, alliances and networks of this kind can be a means of *circumventing politically sensitive issues*, as illustrated by the example of the 'selling of Swedish hydro' noted earlier. Thus carefully designed alliances offer a means of influencing what gets put on policy agendas or raised in the public debate – and, similarly, what stays off these agendas and debates.

Finally, just as managers work to depoliticize their essentially political agendas, it is important to reflexively point to my role as researcher in working to repoliticize them. My agenda is shaped by wanting to question the expressed motives of 'heroic' managers, to identify neglected issues, and to show the inseparability of artefacts, economics and politics.

An unexplored issue concerns the implications of these new alliances for the many 'invisible' or underrepresented actors (see e.g. Star 1991, Fujimura 1991 and Clark and Montini 1993) who are far from corporate negotiating tables but whose everyday lives are influenced by changes in the organization of who lights their homes, with what resources, and on what terms. How will extensive alliances among large utilities impact upon the competitive positions of actors such as smaller, local suppliers? How will they influence the prospects of achieving and sustaining a dispersed system of electricity supply (Summerton and Bradshaw 1991), and how can regulatory influence be insured in ever-expanding systems beyond traditional boundaries? What will be the future channels of influence for small consumers such as you and I: we will know who to call if the lights go out or the fallout comes?

Notes

1 Autonomy is also a salient characteristic of joint ventures: 'they meld people and parts of institutions to the making of a large technical systems, even though each conjoined group maintains their intellectual and strategic independence' (Bugos 1996: 4, and personal communication, 3 May 1996).

2 Operation, ownership, and control of the national trunkline grid had already been transferred from state-owned Vattenfall to an independent state company, the Swedish National Grid, in 1992.

3 Interview with Benny Carlsson, vice-president, Vattenfall International, 19 May 1995.
4 Interview with representative of Vattenfall.
5 A representative from Vattenfall also used this term in an interview, admitting the political volatility of the issue.
6 The hydro which SK Power acquires is reportedly about 1 per cent of Sweden's total hydro production, while VAT's share of the CHP capacity is the equivalent of about 2 per cent of Denmark's total electricity production. See 'Samarbete mellan Vattenfall och Sjaellandske Kraftvaeker ger förutsättningar för stabila och låga priser', 17 August 1993: 1.
7 Interviews with TVL's former chief executive, 9 May 1995 and 17 May 1995.
8 Ibid.
9 Sydkraft's much-publicized acquisition of Malmö Energy Company in April 1991 was one of the earliest examples of this trend.
10 Interviews with TVL's former chief executive, 9 May 1995 and 17 May 1995; interviews with representatives of MSE and TVL.
11 Linköping municipality, planning committee of Executive Countil, minutes from meeting 5 May 1991, § 305, dnr 91.0434.
12 Earlier in the process, politicians on Mjölby's Executive Council had declared that 'the political will was that the municipality should retain its dominant influence in MSE'. See 'Samverkan Mjölby-Svartådalen Energiverk AB – angränsande energibolag', Mjölby municipality, 30 December 1992: 3.
13 Remaining shares would be owned by another municipality which was a pre-existing owner. The proposal also included provisions for buying out Mjölby's pre-existing private shareholders.
14 Conversations with MSE's chief executive and TVL's former chief executive, 9 June 1995.
15 MSE letter to Mjölby Executive Council ('Angående bildandet av råkraftbolag'), 6 April 1992.
16 This figure does not include a company for purchasing natural gas if and when such options are (again) actualized in Sweden. The company, Östgas, is jointly owned by TVL (51 per cent), the Norrköping energy company (35.7 per cent), MSE (6.6 per cent) and Motala municipality (6.6 per cent).
17 Interviews with TVL's former chief executive, 9 May 1995 and 17 May 1995.

References

Annual Reports (1993, 1994) and other materials issued by Vattenfall AB; i/s Sjaellandske Kraftvaerker; Tekniska Verken i Linköping AB; Mjölby-Svartådalen Energiverk AB; Elkraft; and Nordel.

Armstrong, C. and Nelles, H.V. (1986) *Monopoly's Moments: The Organization and Regulation of Canadian Utilities, 1830–1930*, Philadelphia (PA): Temple University Press.

Bugos, G. (1996) 'System reshapes the corporation: joint ventures in the Bay Area Rapid Transit system 1962–1972', paper presented at conference 'The Spread of the Systems Approach', The Dibner Institute, Cambridge (MA) 3–5 May.

Callon, M. (1986a) 'The sociology of an actor-network: the case of the electric vehicle', in M. Callon *et al.* (eds) *Mapping the Dynamics of Science and Technology: Sociology of Science in the Real World*, London: Macmillan.

—— (1986b) 'Some elements of a sociology of translation: domestication of the scallops and the fishermen of St. Brieuc Bay', in J. Law (ed.) *Power, Action and Belief – A New Sociology of Knowledge?*, London: Routledge.

Caron, F. (1988) 'The evolution of the technical system of railways in France from 1832 to 1937', in T.P. Hughes and R. Mayntz (eds) *The Development of Large Technical Systems*, Boulder (CO): Westview.

Clark, A. and Montini, Th. (1993) 'The many faces of RU486: tales of situated knowledges and technological contestations', *Science, Technology and Human Values* 18, 1: 42–78.

ERA (1995) *Elektricitetens rationella användning.*

Euroenergy (1994a) 'Bilateral agreements in future cooperation', autumn.

—— (1994b) 'Electricity trading: putting technical differences to good use', autumn.

—— (1994c) 'EU is good for Vattenfall', autumn.

Fujimura, J. (1991) 'On methods, ontologies, and representations in the sociology of science: where do we stand?', in D. Maines (ed.) *Social Organization and Social Process: Essays in Honor of Anselm Strauss*, New York: de Gruyter.

Hirsch, R.F. (1989) *Technology and Transformation in the American Electric Utility Industry*, Cambridge: Cambridge University Press.

Hughes, T.P. (1983) *Networks of Power: Electrification in Western Society 1880–1930*, Baltimore (MD): Johns Hopkins University Press.

Kaijser, A. (1991) 'Trans-border integration of electricity and gas: the case of the Nordic countries', paper presented at conference 'Trans-Border Aspects of Large Scale Technical Systems', Sydney, 1–6 July.

Latour, B. (1987) *Science in Action: How to Follow Scientists and Engineers through Society*, Cambridge: Cambridge University Press.

—— (1988) 'The prince for machines as well as for machinations', in B. Elliot (ed.) *Technology and Social Process*, Edinburgh: Edinburgh University Press.

Law, J. (1987) 'Technology and heterogeneous engineering: the case of Portuguese maritime expansion', in W. Bijker, T.P. Hughes and T. Pinch (eds) *The Social Construction of Technological Systems: New Directions in the Sociology and History of Technology*, Cambridge (MA): MIT Press.

Midttun, A. (ed.) (1997) *European Electricity Systems in Transition: A Comparative Analysis of Policy and Regulation in Western Europe*, Oxford and New York: Elsevier Science.

Nyhedsmagasinet Sjaellandske Kraftvaerker (1993) 'Samarbejdsaftale mellem SK energi og Vattenfall', nr 2.

—— (1994) 'Vattenfall og SK energi går i taet samarbejde', nr 1.

Robischon, Tobias (1994) 'Transformation through integration: the unification of German telecommunications', in J. Summerton (ed.), *Changing Large Technical Systems*. Boulder (CO): Westview, 119–40.

Salsbury, S. (1988) 'The emergence of an early large-scale technical system: the American railroad network', in R. Mayntz and T.P. Hughes (eds) *The Development of Large Technical Systems*, Boulder (CO): Westview Press.

Singleton, V. and Michael, M. (1993) 'Actor-networks and ambivalence: general practitioners in the UK cervical screening programme', *Social Studies of Science* 23: 227–64.

Star, S.L. (1991) 'Power, technology and the phenomenology of conventions: on being allergic to onions', in J. Law (ed.) *A Sociology of Monsters*, London and New York: Routledge.

Summerton, J. (1992) *District Heating Comes to Town: The Social Shaping of an Energy System*, Linköping, Sweden: Linköping Studies in Arts and Sciences.

Summerton, J. and Bradshaw, T. (1991) 'Towards a dispersed electrical system: challenges to the grid', *Energy Policy* 19, 1: 33–55.

Williamson, O.E. (1990) 'The firm as a nexus of treaties: an introduction', in M. Aoki, B. Gustafsson and O.E. Williamson (eds) *The Firm as a Nexus of Treaties*, London: Sage Publications.

7

FROM CONTROL TO COORDINATION

New governance models for information networks and other large technical systems

Janet Abbate

The rapid growth, interconnection, and convergence of large technical systems raise complex new governance issues. As systems expand in size and scope, public oversight becomes more urgent but also more difficult using traditional policy instruments. This chapter examines ways in which changing systems challenge existing approaches to governance and suggests areas in which new models are needed to guide policy. Current issues in standards policy for information networks illustrate the advantages and pitfalls of some new approaches to government oversight.

Challenges to existing modes of governance and the need for new models

The retreat from direct government control

Historically, national governments and international regulatory bodies and agreements have exerted considerable control over large technical systems. Government has played an important role in the emergence of many such systems by sponsoring research and development (for example, in computing systems), subsidizing commercial enterprises (railroads), or building systems for its own use (highways). Even when systems have been constructed privately, government has asserted the authority to regulate on the grounds that public resources (rights of way, broadcast spectrum, etc.) were involved, that vital national interests such as defense were at stake (atomic energy, data encryption), or that monopolies needed to be checked (telephone, cable).

Since the 1980s, direct state intervention into many large technical systems has been challenged. Government's identity as system builder has become problematic in much of the developed and developing world as political trends such as deregulation, privatization, free trade agreements, and scaled-back budgets for development or defense projects dictate a reduced government role. National telephone, railway, and utility systems are being privatized in many countries, changing government's role from ownership to oversight of private enterprises. Private-sector firms offer services that threaten to compete with or replace public systems, and increasingly international markets have made it harder to isolate state enterprises from competition. Some governments have sold off assets to improve balance sheets, while the end of communist rule in other countries has signaled a reduction in the state's economic role. Under these pressures, governments are increasingly moving from direct control in the form of ownership to coordinating markets and overseeing systems owned by others.

Government has in many cases become a consumer rather than a builder of systems. This is especially striking in the fields of computing and communications, where many systems were first built or funded by defense agencies (radio, microwave communications, various computer systems and networks) but are now supplied by the private sector. As a consumer of off-the-shelf technology, government is spared the cost and risk of investing in new systems, but can affect design decisions only indirectly. In areas where the state continues to represent a significant fraction of the market, governments can remain directly involved by requesting specific technologies; but while procurement has long been one method for exerting government control, attempts to 'pick winners' carry the risk of diverting resources to inefficient or unpopular technologies. If procurement is used to shape emerging technologies, it may be safer to specify 'meta-standards' (discussed below) rather than particular technologies.

In many areas the scope of government control has changed. Instead of operating entire systems, governments may assert regulatory authority over strategic sites (such as system interfaces) or aspects (such as safety or tariffs). Rather than concerning themselves with the internal workings of systems, policy makers are shifting attention to the interactions *between* systems – among competitors in the same industry, between related or complementary industries, across local or national boundaries. This is especially true where formerly separate systems, such as telecommunications and information technology, are converging; the interfaces between these systems often represent boundaries between technical disciplines, economic actors, and political entities.

As more and more systems cross national borders, governments are also realizing that they cannot devise policies in isolation but must coordinate with other nations on issues ranging from tariffs to technical standards to copyright protection to cultural norms such as use of local languages. An example of the problems encountered when governments try to act unilaterally is the United States' unsuccessful attempt in the mid-1990s to impose the 'Clipper Chip' encryption technology on devices used for international communications. The Clipper Chip would have allowed otherwise secure communications to be tapped by US intelligence agencies, which provoked objections by other countries as well as US corporations with multinational operations.

The need to agree on procedures for transnational activities has, of course, long been recognized, but even systems internal to a single country are now exposed to the international market. Multinational corporations and other international organizations can pick and choose among national systems, shifting their operations to the most favorable regulatory environment. A country whose regulations follow regional or international norms will be more attractive to foreign investment; conversely, that country's own enterprises (public or private) will fare better in foreign markets if they do not have to adapt to a multitude of conflicting regulations. In response to the internationalization of markets, purely local control of systems has been giving way to trading and regulatory blocs such as the European Union, Association of South East Asian Nations (ASEAN), North American Free Trade Agreement (NAFTA), and the General Agreement on Tariffs and Trade (GATT), which have begun to include network goods and services in their domains (see McGowan's contribution to this volume).

In response to these changing roles, new approaches to system coordination are being developed that recognize the limits of government authority and seek effective and appropriate alternatives for achieving policy aims. Government can shape the environment for private enterprise through means such as trade policy, standards setting, and patent law. Public agencies can offer services that help coordinate, legitimate, and certify private-sector activities. Governments can also leverage public opinion by publicizing issues involving large systems; a recent example is the publicity surrounding the 'National Information Infrastructure' (in the United States) or the 'Global Information Society' (in the European Union).

Growth and convergence of network systems

Several characteristics of networks pose particular governance challenges. One challenge arises from the convergence of independent systems into interconnected meta-systems, as when telecommunications and computing are combined in new information networks. Governments may be called upon to reconcile differences between systems with entrenched technical methods, work practices, and organizational cultures. Another potentially significant characteristic of networks is their geographic spread. Because the purpose of many network systems is to link people or sites of production or consumption across distances, they tend to have a wide geographic scope even when operating on a small scale economically. Once established, many network systems grow rapidly, not only to achieve economies of scale but also because new users add value to the system. As networks grow they may undergo conflicting pressures toward centralized control, which promises efficiency, and decentralization, which promises responsiveness, flexibility, and local autonomy.

Many networks are characterized by a two-way, user-directed flow of goods or information. This adds to the complexity of these systems, because system uses multiply as the number of users grows. System users tend to be diverse and fill a variety of roles within the system. Describing information networks, Branscomb and Kahin note that

> Users may include resellers, integrators, publishers, value-added networks, libraries, or companies with private networks, as well as individual end-users. Except for the end-users, all of these are also vendors. [. . .] Thus, the vendor/user distinction is of less value for classifying stakeholders than it is for describing relationships within a complex infrastructure.
>
> (Branscomb and Kahin 1995: 10)

In order to ensure that users are active co-creators of the system, policy makers need to allow for user input and strive to maximize system flexibility.

The need for new models

In order for governments to shift their approach from overall control to strategic intervention, they will need new socio-technical models of large technical systems. Models can aid policy by identifying the actors and interests involved and pinpointing technical areas that affect policy goals such as efficiency, coherence, flexibility, and fairness.[1] An emergent networked system creates a new 'space' (physical or virtual) for commerce, speech, and social interaction. This new space requires new technical, legal, regulatory, economic, and political models as well as coordination between historically distinct policy spheres. Questions of jurisdiction must be addressed as expanding systems begin to erode local and national boundaries and straddle the line between public and private.

Regulatory models are one policy area that must be reexamined. New systems challenge old regulatory assumptions by taking on new functions (telephone and television become interchangeable). Systems once thought to be 'natural monopolies' may be deregulated as they become subject to competition (telephone), while new services may invite regulation as they become ubiquitous (cable). As systems that have operated under different regulatory models converge, government must choose or create a model for the resulting meta-system (de Sola Pool 1983).

New economic models are also needed that weigh the trade-offs of government involvement in network systems. Is government subsidy of network systems a distortion of the 'market' or an investment in 'infrastructure'?

Should policy allow or encourage competition in the provision of network goods and services? If so, at what level? Is monopoly more acceptable in the provision of infrastructure than in the provision of value-added services? Is the public good better served if the specifications for widely used technologies are in the public domain, where they can be freely examined and modified, or is innovation better stimulated by the interplay of proprietary, 'black box' technologies? Another issue currently under scrutiny is intellectual property, a legal device that is meant to serve the economic purpose of rewarding innovation. Some economists have argued that network technologies should have less patent and copyright protection, on the grounds that in these systems there is a greater public interest in compatibility (hence shared, rather than proprietary, techniques) as well as a greater risk that firms will reap competitive advantages that are out of proportion with their investment or the value to society of their innovations (see e.g. Farrell 1995, Shurmer and Lea 1995). An appropriate model of the role of intellectual property is needed to guide government agents in interpreting patent and copyright law or negotiating trade agreements.

If network systems become so widely used and depended upon that they are considered part of the basic infrastructure of national life, they take on a new political significance that may demand government involvement. Ubiquitous systems tend to be perceived 'public spaces' where private actors – even the owners of the system – should not be allowed to arbitrarily censor speech or control behavior. If large technical systems are seen as vital economic or social resources, the public may feel that access to these systems should be considered a right and that government should act to extend that right to all, especially disadvantaged groups. In the United States, for instance, local governments often subsidize access to telephone systems for the elderly, and the Clinton administration's plans for a 'national information infrastructure' include extending Internet access to schoolchildren throughout the country. Communications networks are also seen by Americans as a forum for political participation, as shown by radio call-in shows, televised 'town meetings', and email to the White House.

Finally, changes in the nature of systems can challenge the assumptions underlying national and international legal models. For instance, since large meta-systems tend to have numerous participants (often including the users themselves) that provide services or otherwise modify the system, it becomes more difficult to assign liability for damage. Different legal aims may come in conflict: if, in the interest of accountability, system owners are made liable for the actions of users, they will try to protect themselves by demanding knowledge of and control over user activities, which violates user expectations of privacy and freedom. As information networks extend globally, governments must reconcile differences in national or regional laws governing freedom of speech, libel, protection of personal data, copyright, and censorship of offensive or security-related information.

A closer look: governance issues in network standardization

Information network standards provide a useful focus for considering some of the issues outlined here. Standards are a crucial site for government intervention because the choice of standards can affect the level of network interconnection, determine which features and services are likely to be offered, and set the environment for competition between system suppliers. Moreover, standards processes and policies for networks mirror the types of changes taking place in the systems themselves. As systems converge, technical standards must be designed to allow diverse components to interconnect, both 'horizontally' with each other and 'vertically' with complementary or value-added services. These demands require new standard-setting mechanisms that can bridge the gaps between industries, nations, and the public and private sectors.

The standards crisis

The system of standards for information networks rivals the networks themselves in its complexity and illustrates many of the same governance problems. There is a widespread belief among those who participate in developing information technology and communications standards – as well as those who depend on the results – that technological and political changes have strained the process to breaking point. Complaints tend to focus on three points: the overabundance of standards bodies, the slowness of the process, and the inadequacy of the standards produced.

The plethora of standards groups is a problem because it confuses would-be standards users, raises the cost of participating in standards development, and produces multiple competing standards. Dozens of formal and informal standards bodies claim authority to set networking standards at a regional or global level, in addition to the many standards groups operating at a national level. Moreover, the telecommunications and computing sectors have parallel and historically separate standards forums that now claim overlapping jurisdictions over data communications.

The convergence of the information and communications industries, each with its own standards practices and forums, means that participants bring conflicting goals and expectations to the standards development process. For instance, Branscomb and Kahin identify three very different paradigms – based on software, telecommunications, and the Internet – that have been applied to data networks (Branscomb and Kahin 1995: 5–8). For computer software, market competition is intense and the value of interface standards for 'locking in' customers is high, so firms have tried to get or maintain a market advantage by promoting proprietary standards that create distinctions between products and apply only within the product line of a single firm or allied group of firms. The telecommunications world, by contrast, has been characterized by state-owned or state-regulated monopolies and has relied on formal national or international standards that are uniform throughout the integrated national systems and often across the global industry. The Internet illustrates yet a third possibility: a minimal set of universal standards, developed by consensus, is used to hold together an otherwise diverse set of components. Internet standards are designed to provide interoperability (but not uniformity, as in the telecoms world) and to expand consumer choice (rather than restricting it, as in the software world). As this example shows, the very concept of 'standards' is socially constructed and can vary widely. Thus the multiplicity of standards bodies is not merely an artifact of the separate histories of these industries, but also reflects in part a multiplicity of interests that must be reconciled before purely 'technical' standards questions can be addressed.

Many people who depend on data communications standards feel that *the current process for producing standards is too slow*: standards are not ready in time to be incorporated into products, or they are outpaced by fast-moving and unanticipated technological change and become obsolete soon after their introduction.[2] The technical and political trends described above have put pressure on standards bodies to act faster, while also making their task more complicated. Rapidly changing technology means standards are needed faster and must be updated more frequently; growth and convergence of systems means standards from different industries must be reconciled; and the range and number of participants in standards groups have increased.

One reason for the perceived slowness of the process is that as networks become larger, more commercialized, and more publicly visible, standards themselves become more politicized. With more at stake, vendors are less willing to compromise over standards that affect their products. Their representatives may even engage in deliberate delaying tactics, stalling the standards proceedings to let their company's own product (which might be incompatible with the standard under consideration) get ahead (Spring *et al.* 1995: 238). The process is also slowed by lengthy battles over intellectual property rights: if a proposed standard turns out to include intellectual property that the owner is unwilling to license on acceptable terms, the standard may have to be discarded and work begun anew (Shurmer and Lea 1995: 385).

There has also been a higher level of user participation in some standards groups, in part due to efforts by various government bodies to foster the inclusion of social groups such as consumers or educators who have tended to be underrepresented. To understand why increased participation slows down the standards process, it is necessary to know how this process has traditionally worked. Formal or 'traditional' standards bodies are well-established noncommercial organizations with a long history of providing national or international standards. Prominent examples include the International

Standards Organization, International Telecommunications Union, International Electrotechnical Commission, European Telecommunications Standards Institute, and American National Standards Institute. These organizations feature open, voluntary participation and decision making, and have tended to involve a relatively small group of experts who represent producer and sometimes government interests. Though some groups rely on formal voting, especially for the final adoption of a standard, most work is done on a consensus basis, which has been feasible because of their relatively homogeneous and technically-oriented membership.

The introduction of new players has strained that consensus process. Larger groups take longer to reach consensus, and members who are not technical experts (such as consumers) may take considerable time to understand the decisions in question. Increasing user representation can increase the difficulty of reaching consensus in a timely manner because the interests of producers and consumers often conflict: consumers are likely to want low-priced, interchangeable commodities, while producers seek greater profits by differentiating their products and imposing a cost for switching brands. This has led some participants to argue that delay is the inevitable price of a more democratic and representative process. For instance, the president of the American National Standards Institute argued in 1995 that uncontested standards could achieve formal approval by ANSI in a matter of months, while in more controversial cases 'additional time is invested precisely because the system is responsive to the need to reconcile diversity among all directly and materially affected parties' (Mazza 1995: 519).

A common complaint from both vendors and consumers is that even when formal standards are available, *these standards fail to satisfy users' requirements*. Adopting a standard means accepting various costs, including the loss of existing or potential customized features, decreased flexibility for future modifications of the system, and delay (Libicki 1995: 37). If a standard does not seem to be worth these costs, users or vendors may opt for a proprietary technology instead, in which case the standards effort has been wasted. For instance, users often look to public standards to provide interoperability between different vendors' products. This would allow users to substitute one product for another, combine products from multiple vendors, or exchange data across systems. If public standards fail to guarantee interoperability, users may have little incentive to adopt them.[3] Vendors, meanwhile, have found some formal standards too complicated and costly to implement, especially when standards bodies have embarked on large, top-down standards efforts.[4] One example of an unwieldy information technology standard that has struggled to win support is OSI (Open Systems Interconnection), an attempt begun in the mid-1970s to create a comprehensive set of standards for computer networking. OSI has so far failed to displace alternative standards (such as TCP/IP, developed by the US government), both because the specifications took a long time to develop, so that other standards became entrenched, and because the OSI standards were considered by many to be burdensome to implement.

Many people concerned with standardization believe that standards would be more likely to meet user needs if users become more involved in the standards development process. Large organized users such as banks and airlines can often put pressure on vendors to meet their requirements, and powerful groups of users may participate directly in the standards process by creating 'user alliances' to formulate requirements or specifications (Ferné 1995: 457–61; Wagner *et al.* 1995: 183–5). For instance, manufacturers who use computer-aided design software have long been frustrated by their inability to exchange information between different proprietary systems, and vendors have had little incentive to agree on compatibility standards. In response, a group of large manufacturers including Ford, General Motors, and Boeing joined together in 1984 to develop, implement, and promote a set of standards called STEP (Standard for the Exchange of Product Model Data) that provides a common language for representing data and thus allows translation between different proprietary systems (Wagner *et al.* 1995: 186–8). Small users, however, have more difficulty making their voices heard. Participating in standards development is costly in terms of transportation and time and requires a level of technical expertise not possessed by most users. To some extent government agents may try to act on their behalf, but there is no systematic representation for user interests.

New approaches

In response to the perceived crisis in standards development, participants have reshaped existing institutions, created novel institutional structures to coordinate system development, and proposed new models for network standardization and regulation.

New private sector institutions

The 1980s and 1990s saw the rise of industry consortia as an alternative to the traditional standards bodies. The consortium idea gained popularity in the computing industry in the early 1980s, when the realignment of computing technology away from standalone mainframes and toward networked systems created a crisis for the industry: customers were becoming frustrated by the lack of compatibility between computer systems and began buying fewer products. Companies responded by joining together to work out common standards that they hoped would make their products more interoperable and hence more attractive to consumers (Branscomb and Kahin 1995: 11, Cargill 1995: 86). Consortia embody their members' joint interest in establishing and maintaining core standards, and they allow smaller companies to collectively counter the market power of large firms and their proprietary standards (Updegrove 1995: 321). The move to consortia reflects a new competitive strategy: member companies stress compatibility within their industry group and compete with each other on the basis of extra features rather than unique (and incompatible) systems.

Consortia are appealing to industry because they promise faster results and less conflict than the formal standards organizations. Consortia restore a condition the formal bodies have lost – control of the process by a small group of like-minded experts – and thus can achieve consensus more quickly. Consortia tend to be better financed than traditional standards organizations, and corporate management is often more committed to promoting their activities and products. Because they are small and focused, consortia tend to be cost effective, and can fill niche areas unrecognized by larger standards bodies (Isaak 1995: 107, Updegrove 1995: 323). Some promoters of consortia even argue that they help improve traditional standards organizations by providing an example of how to streamline the standards process – and a competitive incentive to do so (Branscomb and Kahin 1995: 12).

However, the trend toward private-sector standards development raises a number of governance issues. Consortia occupy an ambiguous place in the public–private spectrum: while they are usually formed by vendors, they are often non-profit, have open (if expensive) membership, make their standards freely available, and often contribute standards specifications to the formal standards organizations, who may then approve them through their usual process. Given that consortia will probably continue to compete with traditional organizations in setting standards, government could usefully discriminate between consortia, keeping a regulatory eye on those that simply promote proprietary standards and granting a measure of legitimacy to those that provide freely available, industry-wide, competitively neutral standards (Updegrove 1995: 332–3). Government could also play a role in coordinating the activities of consortia to avoid duplication of standards.

The main drawback of consortia from a policy point of view is their lack of due process and wide representation. Government could try to widen representation by subsidizing the participation of user groups, and could offer technical training to public-interest representatives to ensure that their participation enhances rather than impedes standards development. But due process – which includes the balanced representation of stakeholders and the opportunity for public comment – will only be found in the formal standards bodies. If governments do not wish these to be bypassed by commercially-oriented consortia, they will need to address the issues of slowness and overly complex specifications that have caused vendor disenchantment with the traditional process.

Regional organizations

Standards bodies tend to mirror the shape of the systems they govern. As information networks grow or merge to cross national borders, interested government or private-sector groups in adjacent countries may form regional organizations to coordinate technical standards. These new institutions provide an intermediate forum between national and global organizations, allowing regional groups to agree on common standards without requiring world-wide consensus.

The European Union has set up a number of bodies and guidelines for computing and communications standardization, including ETSI (European Telecommunications Standards Institute) and CENELEC (European Committee for Electrotechnical Standardization). In telecommunications, ETSI is joined by representatives from Japan, the United States, and other developed countries in the Global Standards Collaboration, which coordinates contributions from regional standards bodies to the International Telecommunications Union. The Organization for Economic Cooperation and Development (OECD) has set up a committee to make recommendations on information technology standardization, and the Organization of American States has a commission on inter-American telecommunications. Multilateral trade agreements are also being used as a mechanism to reach accord on standards. GATT calls on signatories to follow official international standards and to participate in international standards bodies, and an industry-created consultative committee on telecommunications helps companies in Canada, the United States, and Mexico align standards to comply with NAFTA (Ferné 1995: 463, Reilly 1995: 590). These regional agreements can be an expedient way to establish common standards among neighbors, but if different regions make different technical choices they run the risk of creating large blocs of entrenched and incompatible standards. Whether regional standards will prove to facilitate or impede global standardization remains to be seen.

New models

System builders have created a number of new technical models to cope with the complexity of networks. One example is the 'reference model' or 'reference architecture', a kind of meta-standard that identifies where standards are needed throughout the system without specifying the standards themselves. This reduces the complexity of standards development by breaking the process into two steps: deciding upon the scope and purpose of each standard and determining its actual specifications. Reference models became popular in the mid-1970s, the Open Systems Interconnection (OSI) model being the most famous example. Another type of technical model is the 'standards profile'. This is a set of standards that fulfill all the requirements of a system and are compatible with one another. Like the reference model, the profile makes implicit claims about where standardization is appropriate and necessary throughout the system. The difference is that reference models are anticipatory and aim at creating a single set of standards, while profiles select from existing standards and are intended to identify a coherent set of standards from the profusion available.

'Open systems' is a concept that has become closely identified with networked systems. 'Openness' can mean, among other things, interoperability between competing or complementary products, equal access to the system for service and product providers, and transparent interconnection between systems (Abbate 1995). Its purpose is to allow diverse components from independent, competing vendors to fit together into a coherent system. Thus a crucial requirement for an open system is the existence of publicly controlled and freely available standards for key network interfaces. The call for openness highlights yet another issue raised by the convergence of computing and telecommunications. Historically, telecommunications standards have tended to be 'open' – publicly specified and compatible with a wide range of products – both because of regulatory mandates and because they have been designed for integrated systems with little external competition. Computer standards, on the other hand, have tended to be proprietary, because unique products bring higher profits in a competitive market and because until recently computer systems were not expected to be interconnected (Band 1995: 364). As networking became an integral part of computer use, the computer industry embraced the idea of openness (which has been the impetus behind the formation of many consortia), but many vendors are still strongly committed to maintaining proprietary systems.

Governments may choose to play a mediating role by actively promoting the creation of open information standards through formal standards organizations. Governments can also promote openness through new legal models that reconsider the role of intellectual property in network systems. Since controlling a crucial interface (the hardware connection for a disk drive, a computer operating system, a network protocol) can be tantamount to having a monopoly over the entire system, governments may want to reduce intellectual property protection for network technologies to prevent some firms from dominating the market and to encourage competition and diffusion of services (Farrell 1995, Shurmer and Lea 1995). One way to do this is to decide that patent and copyright protection should not extend to interface specifications (Band 1995: 354). Antitrust measures can also be used to break the grip of proprietary standards, as the US Justice Department has attempted to do to rein in corporate giants like IBM and Microsoft.

Strategies for coordination

How can chaos be avoided as policy groups proliferate? What will keep regional or industry groups from going their own way and producing a myriad of incompatible 'standards'? One possibility that has been suggested is to create a standards development hierarchy, in which focused local groups act as feeders to more inclusive national and international forums. Such hierarchies already exist, but could be extended to include standards actors currently left out, such as industry consortia (see e.g. Isaak 1995: 109). A feeder system would separate the process of formulating technical details for standards from the process of deciding which standards to adopt: experts could work out the former, while representatives of all the interested groups could have a say in the latter. This approach has the advantage of efficiency, and would provide a way to enlist specialized standards groups while maintaining global authority. The drawback is that it restricts the opportunity for wider input into technical decision making. Technical choices often have economic or political ramifications; if the public does not have any input until complete specifications are on the table, it may be too late.

Another option is to develop bilateral or multilateral coordination among the various standards bodies, as some standards organizations already do by exchanging representatives between related committees. This would not solve the question of who has jurisdiction over particular standards questions, but would at least provide the information to allow voluntary cooperation between groups. An example of this approach in the United States is the American National Standards Institute, a nongovernment, nonprofit organization that acts as a clearinghouse for voluntary standards efforts: setting rules for standards development, accrediting standards development organizations, and creating coordinating bodies for special needs.

Since the existence of multiple standards for large systems is probably unavoidable, it may be sensible for governments to adopt policies aimed at managing rather than eliminating this diversity. One possible strategy is to establish meta-standards, such as reference models or profiles, rather than mandating the specific content of standards. Meta-standards would attempt to ensure that the various different standards developed for a single application were mutually compatible. Not all policy makers agree that diversity should be accommodated, however. The European Commission, for instance, has formed guidelines that are meant to promote the adoption of a single, formal standard wherever possible. These include favoring formal standards over commercial standards; favoring Europe-wide and international standards over national standards; choosing a single standard when possible, and requiring interoperability between standards if a single standard cannot be achieved; promoting conformance and interoperability testing; and using public procurement to support formal standards (Bucciarelli 1995: 425–7).

Finally, government can provide facilities for testing standards implementations and certifying standardized products. This would reduce the cost to industry of complying with standards and would provide consumers with the assurance that products provide the interoperability they expect and require. A government facility for standards verification would be seen as unbiased toward any particular vendor and could double as a clearinghouse for information on standardized products.

Conclusion

The case of information network standards illustrates how new governance issues are raised when changes in the structure of network systems – such as a move from public to private ownership, the convergence of telecommunications and computing systems, or the merging of regional systems – introduce new interest groups. Balancing or deciding among these competing interests is often a necessary precursor to addressing more technical questions. Policy makers can help create a balance of interests by establishing hierarchies of authority among different groups or by embracing models such as 'open' meta-standards that attempt to level the playing field.

As network systems grow and converge and trends toward privatization and deregulation accelerate, governance will increasingly take the form of coordination rather than ownership. Direct control of systems will be replaced by indirect means such as regulation, standardization, investment in research, and procurement. While private-sector alternatives to formal policy channels are likely to remain an important force, government leadership is still urgently needed in many areas where the public interest is at stake or where international coordination is needed. Governments can be most effective not by trying to assert authority where it no longer exists, but by focusing on those sites where intervention is most needed and taking advantage of their power to coordinate as well as control.

Notes

1 For examples and theoretical discussion of the relevance of system models for policy see Bijker *et al.* 1987.

2 These issues were discussed in depth at a June 1994 workshop on standards development for information infrastructure sponsored by the National Institute of Standards and Technology, the Science, Technology and Public Policy Program at Harvard University, and the Technology Policy Working Group of the Information Infrastructure Task Force. Most of the representatives of industry, government, and standards organizations who participated agreed that the traditional standards process had become bogged down. There was some disagreement on whether delays were avoidable or an inevitable consequence of the democratic process, and whether traditional standards bodies should be reformed or replaced. For representative discussions of the problem see Ferné, Lehr, Libicki, Spring *et al.*, and Wagner *et al.*

3 In fact, standards makers – who still usually represent industry – may deliberately avoid interoperability, because they feel it will turn their products into commodities and thus lower their profit margins (Lehr 1995: 123).

4 Ironically, these top-down standards efforts were themselves attempts to respond

to the fast pace of technical change and the convergence of diverse systems. By creating anticipatory standards, the standards bodies hoped to keep pace with evolving technology, and by creating comprehensive systems they hoped to provide interoperability. The question of how to devise standards that are neither chaotic nor over-engineered has yet to be resolved.

References

Abbate, Janet (1995) '"Open systems" as a socio-technical model for networks', paper presented at the Annual meeting of the Society for the History of Technology (October).

Band, Jonathan (1995) 'Competing definitions of 'openness' on the National Information Infrastructure', in B. Kahin and J. Abbate (eds) *Standards Policy for Information Infrastructure*, Cambridge (MA): MIT Press, 351–67.

Bijker, Wiebe E., Hughes, Thomas P., and Pinch, Trevor (eds) (1987) *The Social Construction of Technological Systems*, Cambridge (MA): MIT Press.

Branscomb , Lewis M., and Kahin, Brian (1995) 'Standards processes and objectives for the National Information Infrastructure', in B. Kahin and J. Abbate (eds) *Standards Policy for Information Infrastructure*, Cambridge (MA): MIT Press, 3–31.

Bucciarelli, Paola (1995) 'The current debate on IT standardization policy in the European Union', in B. Kahin and J. Abbate (eds) *Standards Policy for Information Infrastructure*, Cambridge (MA): MIT Press, 421–9.

Cargill, Carl F. (1995) 'A five segment model for IT standardization', in B. Kahin and J. Abbate (eds) *Standards Policy for Information Infrastructure*, Cambridge (MA): MIT Press, 79–99.

Farrell, Joseph (1995) 'Arguments for weaker intellectual property protection in network industries', in B. Kahin and J. Abbate (eds) *Standards Policy for Information Infrastructure*, Cambridge (MA): MIT Press, 368–77.

Ferné, Georges (1995) 'Information technology standardization and users: international challenges move the process forward', in B. Kahin and J. Abbate (eds) *Standards Policy for Information Infrastructure*, Cambridge (MA): MIT Press, 455–65.

Isaak, Jim (1995) 'Information infrastructure meta-architecture and cross-industry standardization', in B. Kahin and J. Abbate (eds) *Standards Policy for Information Infrastructure*, Cambridge (MA): MIT Press, 100–18.

Lehr, William (1995) 'Compatibility standards and interoperability: lessons from the Internet', in B. Kahin and J. Abbate (eds) *Standards Policy for Information Infrastructure*, Cambridge (MA): MIT Press, 121–47.

Libicki, Martin C (1995) 'Standards: the rough road to the common byte', in B. Kahin and J. Abbate (eds) *Standards Policy for Information Infrastructure*, Cambridge (MA): MIT Press, 35–78.

Mazza, Sergio (1995) 'The role of ANSI in standards development for the Information Infrastructure', in B. Kahin and J. Abbate (eds) *Standards Policy for Information Infrastructure*, Cambridge (MA): MIT Press, 516–30.

Reilly, Arthur K (1995) 'Defining the US telecommunications network of the future', in B. Kahin and J. Abbate (eds) *Standards Policy for Information Infrastructure*, Cambridge (MA): MIT Press, 579–93.

Shurmer, Mark, and Lea, Gary (1995) 'Telecommunications Standardization and Intellectual Property Rights: A Fundamental Dilemma?', in B. Kahin and J. Abbate (eds) *Standards Policy for Information Infrastructure*, Cambridge (MA): MIT Press, 378–402.

Sola Pool, Ithiel de (1983) *Technologies of Freedom*, Cambridge (MA): Harvard University Press.

Spring, Michael B., Grisham, C., O'Donnell, J., Skogseid, I., Snow, A., Tarr, G., and Wang, P. (1995) 'Improving the standardization process: working with bulldogs and turtles', in B. Kahin and J. Abbate (eds) *Standards Policy for Information Infrastructure*, Cambridge (MA): MIT Press, 220–50.

Updegrove, Andrew (1995) 'Consortia and the role of the government in standard setting', in B. Kahin and J. Abbate (eds) *Standards Policy for Information Infrastructure*, Cambridge (MA): MIT Press, 321–48.

Wagner, Caroline S., Cargill, Carl F., and Slomovic, Anna (1995) 'Open systems standards in manufacturing: implications for the National Information Infrastructure', in B. Kahin and J. Abbate (eds) *Standards Policy for Information Infrastructure*, Cambridge (MA): MIT Press, 178–97.

8

THE INTERNATIONALIZATION OF LARGE TECHNICAL SYSTEMS

Dynamics of change and challenges to regulation in electricity and telecommunications

Francis McGowan

Introduction

This chapter examines the regulatory implications of the internationalization of the large technical systems (LTSs) which deliver utility services in the energy and telecommunications fields. Changes in the economic, political and technological context have ushered in a wave of restructuring and reform in these industries, in many cases undermining their position as 'national champions'. At the same time, the traditional forms of regulation over the international activities of LTSs are undergoing radical changes. The industries themselves are 'internationalizing' as utilities in the developed and developing worlds look abroad for market opportunities through joint ventures, take-overs or direct competition. The existing sectoral and intergovernmental agencies which managed international transactions in the past are having to adjust to this changing environment while other international agencies, regional and global, in the fields of trade liberalization and finance are effectively speeding the pace of reform. Yet the scale of the changes leaves many issues unresolved, most importantly the regulatory environment within which internationalized LTSs operate: if firms and markets transcend national boundaries what are the prospects for regulation? Put another way, is effective international regulation possible?

This chapter reviews the process of internationalization in LTSs and considers the regulatory implications of the changes. It begins by noting why the changes are taking place – focusing on the regulatory structures which have prevailed in these sectors over many decades – and then turns to examine the emerging regulatory problems of this new system. After discussing why international regulation might be necessary, the chapter outlines the international regulatory landscape, contrasting the traditional sectoral regimes with the emerging regional and global regimes. The chapter concludes by assessing whether the shift in international regulation is able to cope with internationalized LTSs. Before considering the process of internationalization and regulation it might be useful to consider what is understood by these concepts here.

The concept of internationalization is used rather than the currently fashionable term globalization.[1] Too often the latter is used as if it were an accomplished and irreversible fact, whereas we would want to argue for a more tentative view of the trends under way. Moreover, 'internationalization' allows us to capture a much wider range of phenomena, most notably the various experiments in regional integration in the developed and developing world. Our interest in internationalization is primarily with institutional rather than operational developments: it is the firms that operate LTSs, the structures of markets they create and the activities of governments and international agencies which concern us rather than the technology of internationalization of LTSs itself.

The emphasis on regulation reinforces the institutional concerns of the chapter. LTSs studied here have presented a range of regulatory challenges including those of monopoly and access, standardization, safety and environmental protection. By regulation we refer to the actions of formal regulatory organizations and the interaction of government and the LTSs more generally. Such regulation has been carried out within national boundaries with utilities in different countries for the most part respecting each other's particular agendas and liaising on a cooperative basis internationally. The extent of international regulation has generally been modest, and has been mainly focused on the activities of sectoral intergovernmental organizations such as the International Telecommunications Union or of industrial associations such as Unipede. In this chapter we touch on both dimensions as well as the wider range of international agencies and arrangements (at both the regional and the global levels) which have become involved in the LTS industries over the last decade or so.

From internationalization to nationalization and back again?

Towards national champions

Although this chapter is concerned with the internationalization of LTSs as a contemporary development, it is as well to recall that, in terms of their history, these industries were very international in character. The concentration of expertise, intellectual property and capital in the nineteenth and early twentieth centuries meant that the initial development of many national LTSs was achieved with the participation – even the leadership – of outside companies and investors. In this section we review the development of LTSs under foreign tutelage, analyse the reasons for 'nationalization' and the successes and failures of this change, and consider the contemporary pressures for internationalization.

British and American firms were heavily involved in the development of utility services in the nineteenth and early twentieth centuries. British firms were active in providing town gas supplies in German cities in the nineteenth century,[2] the leaders in the electricity industry such as Edison and Westinghouse were involved in the development of supply in much of Europe.[3] American telephone companies were responsible for telephone and telegraph services in a number of European countries.[4] While it is true that the larger European powers were either in the forefront of development or rapidly established control over these industries (either through public ownership or the participation of local capital) in some countries, foreign control of LTSs persisted in some form into the 1960s. For developing countries, foreign investment played an even more important role in the development of these industries. From the late nineteenth century investments in developing countries were flowing from a variety of developed countries, including the UK and Canada, though in the 1920s and 1930s the primary investors were large US corporations and holding companies. These had expanded primarily in the US domestic market and looked abroad to diversify, often taking over from other foreign investors in Latin America and Asia (Wilkins 1970, 1974).

From the 1940s on, however, the pattern began to change. In Europe, domestic capital became increasingly predominant in the utility industries, a process confirmed by the postwar nationalizations in Britain and France. Within many developing countries, particularly Latin America, the momentum of foreign investment faltered, due to three factors: depreciating currencies, constraints on profit repatriation and controls on rate making (Wilkins 1974, Tendler 1968, Wionczeck 1964, Cavers and Nelson 1959). Rate increases did not keep pace with foreign exchange losses and relations with host governments deteriorated. Foreign owners cut their investment in local systems compounding performance problems and heightening tensions with governments which were becoming emboldened by a new wave of economic nationalism. Ironically their stances were reinforced by the radical changes under way in the developed world. The postwar nationalizations were an important exemplar for many former colonies while the US government's creation of a major publicly owned utility in the shape of the Tennessee Valley Authority was a particularly resonant example for many Latin American countries which were seeking to exploit large water resources. Thus as US and other foreign investors' interest declined, divestment and, in many cases, nationalization followed in the 1950s and 1960s (Wilkins 1974).[5]

The telecommunications sector followed a similar path of nationalization (Kiplagat and Werner 1994).

Given their nature it is perhaps not surprising that LTSs became both symbols and instruments of independent, autonomous economic activity, central to the economic nationalism of Europe in the interwar and immediate postwar period and of the developing world in the following decades. The importance of these sectors cannot be overestimated; as basic infrastructures and strategic industries, utilities have played a central role in economic planning and development. In addition to the basic tasks of meeting demand (with a strong tendency towards 'self-sufficiency' and an emphasis on supply security), such goals as rural electrification, large-scale exploitation of natural resources, the development of supplier industries (or local content requirements in new investments) were also part of the development of the energy and telecommunications LTSs.

This shift in the ownership structure of LTSs can therefore be seen as a result of national regulatory strategies. The regulation of utilities and of foreign investment effectively made foreign investment less tenable in many states. In the case of some developed countries, moreover, national regimes worked against foreign expansion: nationalized firms were often constrained in their investment options to the domestic market while even private utilities found their ability to engage in foreign ventures constrained by national legislation.[6] Moreover, national regulatory strategies – always hinging on the recognition of monopoly and the application of cost plus principles and often mediated through public ownership – were aligned to the wider public policy objectives noted above. There was in effect a match of technological, territorial, political, and economic factors in the regulation of LTSs in this period.

Crisis and critique

The period of nationally-based LTSs reached its peak in the early 1970s. Until then, technological advance in most LTSs was still quite rapid and worked to reinforce existing structures of ownership and control. Moreover, it is arguable that the use of these sectors for national development was most effective in this period. In Europe over the postwar period and in the developing world in the 1960s and 1970s these industries were at the centre of major infrastructural programmes (McGowan 1993; see also World Bank 1994). In the case of the developing countries, these programmes were often backed up by substantial funds from development banks and technical assistance from metropolitan utilities (indeed training and technology transfer constituted the main foreign activities of developed country utilities).

In both developed and developing countries, the economic problems of the 1970s ushered in a period of stagnation and even deterioration in the performance of many LTSs. The nature and scope of such problems varied from country to country and sector to sector though there were some common themes. In developed countries, the virtuous circle of falling costs, rising demand and more investment came unstuck, most notably in the energy sector (where although rising costs of supply slowed demand growth, overambitious forecasts led to over investment and surplus capacity). In developing countries rapid demand growth, allied with increased investment requirements (often on the basis of deteriorating capital stock), financial shortfalls (through controls on prices, cross subsidies and governments taxing or otherwise accruing revenues) and declines in productivity (through overmanning) undermined the performance of many utilities (World Bank 1994). For many critics of the status quo, the basic problem in these sectors was one of regulatory failure arising from the nature of government–industry relations and the monopolistic market structures of most LTSs. From being instruments of economic development, public utilities became symbolic of the broader economic problems facing developed and developing countries. Similarly, in the climate of economic reform in the 1980s and 1990s, they became targets for a new relationship with government. Just as public utilities were central to previous economic strategies, so they became an important component in the reforms of the last ten years.

Beyond the need to rectify the immediate problems with performance in many countries, the pressures for reform included financial constraints, pressures from large users and changing policy preferences of politicians, bureaucrats and even industry management. Moreover, once reform took place in some countries, there was an international 'spillover' or convergence as the experience was emulated in other countries and endorsed by a range of international organizations and as reformed utilities sought access to other markets.[7]

Reform and internationalization

The process of reform in developed and developing countries has focused upon the two aspects of the regulatory failure in utilities, ownership and market structures. The question of ownership has been addressed through policies ranging from 'corporatisation' (Nellis 1988) – the establishment of the utilities as quasi-private corporations which are expected to operate as such in the context of an agreed relationship, or contract, with the government – through to outright privatization. If anything the emphasis has been on the latter as financial constraints have prompted governments to seek private participation. This could take the form of private capital to meet new investment requirements, or of the establishment of a partnership under which private firms would carry out some aspects of utility operations, or it could mean a partial or complete transfer of the utility itself to the private sector (Ramamurti and Vernon 1991, Vuylstecke 1988). In many countries, particularly in telecommunications, there has also been a process of liberalization

Table 8.1 Internationalization in the electricity supply industry

United Kingdom

- Takeovers of privatized Regional Electricity Companies in England and Wales by American utilities/holding companies (e.g. Southern acquisition of SWEB, Avon takeover of EMEB)
- US utilities/IPP participation in British IPP projects (e.g. Enron in Teeside)
- US consortium acquisition of National Grid pumped storage plant

Sweden

- Preussenelektra stakes in Gullspanskraft
- Statkraft stake in Sydkraft
- IVO stake in Gullspanskraft

Portugal

- British utility-led consortia in IPPs (National Power in Pego, PowerGen in Tapada)

Spain

- National Power, EdF and ENEL in Elcogas clean coal project
- RWE stake in Endesa

Germany

- PowerGen in IPP (Schopkau)

Hungary

- EdF stakes in distribution companies

Australia

- PowerGen in production and distribution companies in Victoria

New Zealand

- National Power in IPP Taranaki
- Transalta in distribution company

Peru

- Virginia Power stakes in production and distribution utilities

Argentina

- EdF, Endesa stakes in Edelnor

Brazil

- EdF, AES stakes in Light

China

- Hopewell and AES in IPP projects

India

- Enron, AES in IPP projects
- PowerGen-led consortium in IPP project

Cote d'Ivoire

- EdF performance contract

Guinea

- EdF performance contract

Philippines

- Hopewell stakes in Production

Source: *Financial Times* (various issues, 1994, 1995).

Table 8.2 Internationalization in the telecommunications industry

France
- BT subsidiary established
- France Telecom and Deutsche Telekom and Sprint form Global One
- Bell South partner in Société Française du Radiotélé Mobile Network

Germany
- Cable & Wireless stake in Mannesman Cellular Network

Sweden
- C&W in Tele-2 (rival PSTN)
- BT, Telenor and Tele Danmark set up Telenordia

United Kingdom
- ATT VAN subsidiary

Ireland
- KPN and Telia stake in Telecom Eirrean

Finland
- Telia stake in Telivo

Spain
- Telefonica, Telia, KPN and Swiss Telecom and AT&T in alliance – WorldPartners

Argentina
- Telefonica in Telco Sud
- Stet in Telco Norte

Jamaica
- C&W in Telecommunications of Jamaica

Mexico
- France Telecom, AT&T in Telmex
- Bell South in Comcel

Venezuela
- AT&T and Telefonica in CANTV

Sierra Leone
- C&W in Sierra Leone External Telecommunications

Indonesia
- Deutsche Telekom stake in Sateldino

Source: *Financial Times* (various issues, 1994, 1995).

which has increased opportunities for market access. In these increasingly competitive environments utilities have often opted to develop new services as joint ventures, and international alliances have become commonplace (Noam 1992, Curwen 1995).

The search for new markets by developed country utilities through the diversification and internationalization of their basic activities, and the need for new investment by developing country utilities converge in a re-internationalization of these industries.

In the energy utility sector (see Table 8.1), reforms have been extensive, ranging from cross-shareholdings or takeovers to outright privatization of utilities, and introducing various forms (and degrees) of liberalization.[8] In the telecommunications sector (see Table 8.2), developed countries have engaged in a widespread programme of market liberalization at both a national and a regional level.[9] It is this which has provided the main catalyst for internationalization. Within developing countries, this process is, if anything, further advanced.

The international regulation of international LTSs – basic principles

The paradox of sovereignty

The internationalization (or re-internationalization) of LTSs is thus well under way. However, it is at an early stage (the first significant moves took place in the mid-1980s), and there have been a number of setbacks. Moreover the activity is often concentrated in some countries (whether as hunters or hunted) while others, notably France, appear less willing to open up their own markets (though perfectly happy to take advantage of open markets elsewhere). None the less, changes have taken place and the pace of transformation shows no signs of abating. However, while the changes in utility attitudes, conduct and structures associated with this shift raise a number of regulatory problems, the development of effective regulation lags behind the transformation of firms and markets. While it is true that an important corollary to the reform of utilities has been the formalization of regulatory structures,[10] in most cases these remain nationally based and not necessarily able to cope with the international dimension. The old match of territory, industry structure and political control is less tenable and in some cases unsustainable, particularly as far as the pursuit of wider policy objectives is concerned.

Yet for the most part most governments cling to the notion of nationally based regulation, even as the national basis of LTSs is eroded along with the ability to control them. This adherence to national regulation reflects the wider paradox of governments seeking to retain the trappings of formal sovereignty in decision making at the same time as their real sovereignty is eroded. While there are developments in the international regulatory domain, the slow progress being made is partly a result of this failure of governments to look beyond the defence of their own political sovereignty and act collectively to counterbalance the internationalizing corporate and market trends. Beyond this problem, moreover, lie the very real problems of the division of regulatory responsibilities and the identification of what is to be regulated. Before looking at how the international regulation of LTSs is developing and might develop, it might be useful to consider why international regulation is necessary.

Why international regulation?

The introduction noted the likely scope of regulation. Essentially, regulation is required to rectify various types of market failure, notably questions of monopoly, standards and externalities, as well as concerns of economic and social development (see Kay and Vickers 1988). These problems are particularly acute in the LTSs under discussion here, though that of monopoly is probably sharpest, especially if there is a shift away from public ownership (in itself a form of regulation) towards private ownership and liberalization. In such circumstances we should not be surprised if there should be some effort to keep an eye on those providing the services and on the terms and conditions which they enjoy.

Of course some would argue that regulation is likely to be as much of a can of worms as nationalization and government control and that it would consequently be better to leave things to the market (this view has its adherents in both developed and developing country contexts). Yet one can hardly argue that the reasons for regulation have disappeared in the new context of the 1990s. If anything the reasons are much stronger now than they have been in the past: consider for example the problem of environmental externalities, social and spatial cohesion. The question is where such regulation should take place. One could argue that if everyone is now concerned with regulation then there should be no need for any international regulation. However that presupposes that everybody regulates the same things in the same way. In practice, relations between different national regulatory regimes may be rather harder to reconcile.

Drawing upon the principle of subsidiarity,[11] it might be argued that regulation is best done at the most local level, whether national or subnational. Such authorities will be better informed than higher levels of government about the needs and preferences of those citizens who are affected by the market failure the regulatory intervention is designed to redress. As a general rule the quality of such information is likely to be diminished by distance of the regulators from the locality regulated. Moreover, it could be argued that these authorities are also more directly accountable to their citizens, and therefore under greater pressure to ensure that regulatory decisions are shaped by their needs and preferences rather than the whims or interests of the regulators themselves. This accountability may be a feature of the political system, which gives citizens a voice in their own affairs: the presumption in favour of such regulation arises because national regulators are accountable to many citizens other than the ones directly implicated in a particular decision, and the influence of the preferences of the affected citizens on them may well be swamped by national trends. However, such criteria may not always be effective. The issue to be regulated may not match the powers of national regulators. There may be a temptation on the part of national regulators to meet the needs of those most vocal rather than the wider interest (consumers being more diffuse and less directly interested in many instances than the regulated firms themselves).

It might be argued that this implies that national regulators can coordinate activities among themselves, without resort to any international level of control. If states could simply agree to modify their domestic regulatory policies to achieve a cooperative outcome, such an outcome could be implemented by national governments without any need for delegation of the relevant powers. But in fact there are two reasons why cooperative outcomes, even if desirable in principle, may not be feasible in the absence of delegation. One is the problem of coordination through negotiations between countries, which entails high transaction costs. The other is the problem of credibility in enforcing regulatory agreements once they are reached. This problem is twofold: first, states have an incentive to breach international agreements for short term advantages; second, it may be difficult for parties to verify whether agreements are being kept or not. Some form of international delegation of powers may thus be required to deal with the challenges raised by more internationalized utility markets, and to regulate the conduct of both firms and governments within those markets.

Needless to say, the necessity of international regulation becomes even more acute if the sectors themselves are becoming more internationalized. It is clear that only a few states have the economic and political clout to take unilateral action in curbing the activities of multinational LTSs (though, if they are in that position, they probably have important interests to promote and protect). Aggressive regulation of the sort seen in the 1940s and 1950s in Latin America would meet with a very different response nowadays (as the response of many investors to regulatory reviews in the UK has shown). In many cases, collective action by many states may be necessary to curb effectively the conduct of such utilities. However, the prospects for such action depend on a shared perception of the purposes of regulation. Is it designed to regulate markets and firms for the benefit of the consumer, perhaps limiting the scope for the major international firms in these sectors to abuse their market power? Or is it designed to facilitate internationalization and protect investors, by opening up markets and constraining the ability of governments and national regulators to protect their home markets? Differences in perception – and interest – may make shared action difficult.

The international regulation of international LTSs: a change of regime?

While the problems of international regulation are considerable, changes are taking place in the international regulatory landscape. The balance of purposes served by international regulation is shifting as is the balance of power and influence between different international regulatory regimes (Krasner 1983). We can identify three different types of regulatory regimes with different sources of legitimacy, different priorities and different impacts: the sectoral regime, which is normally industry-based and largely concerned with a range of 'technical' issues, not least those related to facilitating international cooperation and coordination; the regional regime, where LTS questions arise in a much broader framework of economic integration; and the global regime where again a broader range of issues to some extent shapes the perspective on LTSs but where the impact is largely indirect. It could be argued that whereas in the past the sectoral regime has predominated, the regional and global regimes are becoming a more important basis of regulation. While the traditional sectoral regimes were based on the presumption that, as national industries, the international aspect of regulation would be limited, the new regimes are based on a wider logic of market liberalization and redefine the purpose of regulation accordingly.

Sectoral regulatory regimes

The sectoral regulation of LTSs dates back in some cases to the last century, though institutional development has varied according to sectoral specificities and wider political and economic changes (Murphy 1994, Burley 1993). As LTSs developed and the question of international operations arose, agreements amongst governments and/or LTSs were devised at a bilateral and multilateral level, with the latter sometimes embodied in an intergovernmental organization which served as a forum for debating standards and trade on the basis of coordination and cooperation, as a body to articulate the shared interests of that sector and as a means of sharing ideas on 'best practice' within the sector. In other cases, more specifically sectoral organizations have developed, sometimes with a more limited membership than the intergovernmental organizations.

The intergovernmental regime is best illustrated by the International Telecommunications Union (originally the International Telegraph Union). This body was established in 1865 by a number of governments and its membership has grown to include almost all nation states. The intergovernmental structure is hardly surprising given the predominance of state owned PTTs for much of the history of the sector, though it is worth noting that the ITU has also made provision for membership by private organizations (including US telecommunications companies). The organization's regulatory role has been strongest in the area of standards though it has also underpinned the system of market sharing which has predominated in the sector in the past. Indeed the ITU has been viewed as legitimating the structure and conduct of the telecommunications sector (Codding 1991, Jacobson 1973, Krasner 1991).

Intergovernmental organizations are not so common in the energy sector. The principal United Nations agency is the International Atomic Energy Agency, a body which has played an important regulatory role with regard to nuclear safety and nonproliferation (Scheinmann 1973). For the most part the model in the energy sector – most clearly in the electricity sector – is of international industrial associations discussing technical matters and playing some representational roles. In the electricity sector, Unipede and Cigre are largely concerned with technical and to a lesser extent representational issues rather than market allocation. This is done on a bilateral basis with the framework of international trade and transfers managed through regional organizations such as the UCPTE and Nordel (Bruppacher 1988).

What is interesting about most of these arrangements is that they have been developed in the context of national LTSs enjoying largely monopolistic positions within their home markets and seeking to manage the international aspects of their activities. It goes without saying that such arrangements are placed under considerable pressure in a context where the former constituents of the organizations, national utilities, are changing their organizational structures and priorities, and where a mixture of new competitors and foreign investment undermine the national basis of such organizations. In effect the breakdown of the domestic monopolies and the growth of international competition makes it hard to sustain cooperative international arrangements. From being at the forefront of LTSs international activities, these bodies are now lagging behind in the process of internationalization and are in danger of losing their traditional pre-eminence to other organizations (Cowhey 1990).

Regional regulatory regimes

Arguably one source of pressure for this reorganization of LTSs is the regional economic groupings which seem to be undergoing some revival at the moment. The regional example is probably best illustrated by the EU. There are of course other organizations – NAFTA, ASEAN, EFTA, MERCOSUR, etc. – but these have not so far involved the same intensity of economic and political integration as the European Union. The experience of over forty years of cooperation, particularly in the economic sphere, has created a wide range of regulatory responsibilities for the institutions of the Union, particularly the Commission. The Commission's regulatory role has developed in the wake of the Single Market initiative in the 1980s and is best illustrated by its increased willingness to use competition policy to open up previously protected industries, most notably the utilities. In these industries the

Commission has been involved in setting basic rules of conduct in the market place (keeping the playing field level) and ensuring that those rules are implemented effectively. That task has been a difficult one – and highly sensitive politically – but overall it appears to be carried out reasonably effectively. The 1988 Green Paper on Telecoms liberalization, and the follow-up directives have together established a new regulatory regime for the European telecoms sector while it is possible that the recently agreed directives on liberalizing electricity and gas markets could play a similar role in the energy sector. While the less ambitious objectives of other regional trading blocs may render it more difficult to develop regional regimes for regulating LTSs, there is the potential.

This process of regional integration is rooted in the liberalization not only of national boundaries, but also of the regulatory frameworks which protected most national LTSs. Moreover it could be argued that this process of integration is reinforced by the changing dynamics of the markets and corporate structures. It may even spill over into the 'hinterland' of the regional agreement (and here the experience of first EFTA and then Eastern Europe *vis-à-vis* the EU is particularly important). Note also that this process of change is driven by a much broader concern than the sectoral particularities of LTSs. The position of LTSs, in other words, is relatively weaker than in the case of sectoral organizations (though they may retain considerable residual influence on account of their incumbent status). Even so the role of regional economic organizations is by definition regional. Therefore, while these various blocs are clearly of greater economic significance for LTSs than their national component parts, to some extent they face a similar problem of regulatory effectiveness where the LTSs in question are becoming truly global. In such cases, it might be argued, global regulatory regimes are required.

Global regulatory regimes

What can we learn from existing global organizations? We have already noted the mismatch between sectoral organizations and the internationalization of LTSs which is currently under way. There are other global organizations which may impinge upon LTSs, however. Indeed one could argue that these organizations have already had a significant impact.

The most important of these are the international financial institutions, most notably the World Bank. This organization has been at the forefront of the debate on reorganizing state–market relations in the developing world, and LTSs have been a particular concern. Interestingly, in the past the World Bank had supported the development of LTSs within the national, monopolistic framework. It has only been in the last ten years or so that it has challenged the existing organizational and ownership structures of LTSs. Its power partly lies in the sheer intellectual weight of its arguments – it has been to the fore in fostering the 'new conventional wisdom' on regulating utilities.

However its financing role, though relatively small compared with other international lenders, is significant and in recent years, loans have come with strings attached.[12]

The World Bank has focused upon the need for internal reorganization and reform of LTSs, even if a key part of its prescription is an opening-up of the economy to foreign trade and investment. The global forum for trade issues has of course been the General Agreement on Tariffs and Trade (GATT) and more recently the World Trade Organization (WTO). For the most part however, international trade regulation has steered away from the LTSs' territory, focusing instead on manufacturing and agriculture. To some extent the special status and the inward focus of LTSs was respected (or perceived as too difficult and sensitive to tackle).[13]

More recently however LTSs have come into the scope of international trade regulation. As traditional barriers to trade such as tariffs have been tackled in the GATT, so the focus has spread to other sorts of barriers and other sectors. The decision to tackle trade in services within the Uruguay Round of trade negotiations brought in such sectors as telecommunications into GATT arguments.[14] Although the results of those negotiations was relatively limited as regards LTSs, it established a precedent for future activities. Moreover, the extension of international trade regulation in a WTO context brings LTSs under closer inspection: the decision to address competition policy issues brings into question the whole issue of monopoly and its regulation in an international trade context. As this debate unfolds it is highly likely to impinge upon LTSs (Holmes *et al.* 1996).

As indicated the role of these global organizations may be as much about reinforcing the trends towards national reform and liberalization as about setting new international rules. It is not clear whether the WTO will be involved in energy market liberalization (particularly where this tends to be primarily regional in character). However, other international agreements which go beyond traditional regionalism may be of some importance. While the exact significance of it has yet to be tested, the European Energy Charter, designed to promote trade and investment in the energy markets of Eastern Europe and the former Soviet Union and with a membership which extends beyond Europe into Asia and, possibly, America, could apply international trade principles to the energy sector (Dore 1994).

Prospects for international LTSs regulation

The last few years have seen considerable activity in the field of international regulation. However the actual results of such efforts are not so apparent. With the exception of some developments in regional regulation, the mechanisms for effective regulation are not in place. Attempts to develop global regulation have so far been unsuccessful. Talks within the WTO on telecoms services were extended once it was clear that the deadline of Spring

1996 would not be met (Holmes *et al.* 1996). The detailed follow-up to the European Energy Charter has only begun in the wake of the much-delayed ratification process. Attempts to apply broad principles of economic regulation as envisaged in proposals for international anti-trust rules (and which could be applied *inter alia* to the utility industries) remain grounded by differences between governments (Holmes *et al.* 1996). At the same time, the role of sectoral regulatory bodies appears to be in limbo, undermined perhaps by the trends in regional and global regulation but with their precise role and future unresolved.

The resulting regulatory gap might be puzzling to some. After all, the application of a liberal trade logic to LTSs would in many ways confirm the trends already taking place in many countries. Removing barriers to entry and investment in LTS markets would underpin and extend the developments which were discussed earlier in this chapter. However, it should be noted that although international trade law principles are supposedly upheld by all parties, in practice they have been relatively weak and difficult to implement (Downs *et al.* 1996). Moreover attempts to regulate other problems on a multilateral basis have been rather disappointing. The moves to tackle global environmental problems indicate the problems in securing the agreement of countries with very divergent interests (Spinz and Vaahtoranta 1994, Grubb 1992).

It is not at all clear whether multilateral agreements can address the range of regulatory problems that might arise in a world of internationalizing LTSs. It is clear that most of the significant moves towards providing international rules have been of a rather negative form; i.e. they are mainly concerned with removing barriers to entry and/or trade. Moreover the underlying focus on economic efficiency – conceived for the most part rather narrowly – makes it hard to take into account broader questions of social and economic development which continue to affect developed countries (e.g. over new information services) as they do developing countries. In the past the latter concerns have been largely tackled by national authorities and/or utilities providing them from the basis of a protected monopoly. As monopolies tumble and the national focus of firms diminishes, it will be correspondingly harder for national regulators to impose such obligations. It might be argued that the solution is to incorporate such obligations into whatever international regime is developed. However, it is clear that however difficult it is to secure agreement on removing market obstacles, it is even more difficult to agree on 'positive' measures of the sort implied here. The experience in a European context of marrying broader obligations to a fundamentally deregulatory regime is salutary in this respect (see Stoffaës 1995). As noted earlier, the purposes of regulation – the interests as well as the objectives served – may not be easy to reconcile in an international setting.

Conclusion

There are significant changes taking place in the utility LTSs, changes which undermine the national basis of these industries (which has been predominant for more than fifty years). That these changes are not matched by any equivalent shift in regulatory power raises serious questions about the consequences of internationalization but should not be too surprising. The reluctance of governments to pool sovereignty in most cases and the very different interests behind the processes of internationalization mean that even if the principle of international regulation could be agreed upon, defining the content of that regulation would present a formidable challenge. Such international regulation as exists is in a condition of some turmoil as the shift away from the sectorally-based mix of intergovernmental and industrial regimes is replaced by regionally and globally based regimes of liberalization. In some cases regional regulation appears to have been relatively successful. The EU experience has been at one level quite positive though there remain problems of reconciling conflicting policy objectives. However, in other regional arrangements, and certainly in global settings, there remains a considerable regulatory gap. The failure to close that gap would have serious consequences for the accountability of the increasingly international utility LTSs.

Notes

1 The globalizers include those celebrating the trend (see for example the management literature of Ohmae 1992) and those bemoaning it (Giddens 1994, Jones 1995). For more sceptical views of globalization, see Hirst and Thompson (1996) and Krugman (1994).
2 Harbutt Dawson (1912) draws attention to the strong position of English companies in this field in contrast to both water and electricity.
3 On the role of these companies in the early development of the European industry: see Hannah (1979) on the UK, Lanthier (1989) on France, Zamagni (1993) and Webster (1975) on Italy, and Harrison (1978) on Spain.
4 On the role of ITT, see Sobel (1982) and Sampson (1973).
5 Note Tendler's scepticism about her critique of Brazilian hydro development.
6 The US Public Utility Holding Company Act of 1935 (PUHCA) certainly made foreign investment less attractive (see Moeller 1993).
7 On the idea of policy convergence, see Bennett (1991). For accounts of why such policies spread, see Henderson (1989), Ikenberry (1990) and Williamson (1994).
8 Moves to liberalize energy markets have tended to focus on increased scope for independent power projects though in some regions there are plans to liberalize markets across boundaries.
9 See Curwen (1995) on EU, Hufbauer and Schott (1993) on NAFTA and Kurisaki (1995) on global developments.
10 On developed countries, especially Europe, see OECD (1992), Majone (1994) and McGowan and Wallace (1997). On developing countries see World Bank (1994).

11 This section draws upon McGowan and Seabright (1995).
12 Babai (1988), Bierstecker (1990) and Mosley (1989) offer contrasting views of the World Bank's influence. On World Bank lending policy see Stewart-Smith (1995).
13 On the development of the international trading system see Gardner (1969) Jackson (1989) and Hoekman and Kostecki (1995).
14 Drake and Nicolaidis (1992) provide an interesting perspective on the development of the trade in services debate.

References

Babai, D. (1988) 'The World Bank and the IMF: Rolling back the state or backing its role', in R. Vernon (ed.) *The Promise of Privatisation*, Washington: Council for Foreign Relations.

Bennett, C. (1991) 'What is policy convergence and what causes it?', *British Journal of Political Science* 21, 2: 215–34.

Bierstecker, T. (1990) 'Reducing the role of the state in the economy: a conceptual exploration of IMF and World Bank prescriptions', *International Studies Quarterly* 34, 4: 477–92.

Burley, A. (1993) 'Regulating the world', in J. Ruggie (ed.) *Multilateralism Matters*, New York: Columbia University Press.

Bruppacher, F. (1988) 'How European electricity trade is organised', Paper to *FT* World Electricity Conference.

Burnell, J. (1986) *Economic Nationalizm in the Third World*, Brighton: Harvester.

Cavers, D. and Nelson, J. (1959) *Electric Power Regulation in Latin America*, Baltimore (MD): Johns Hopkins University Press.

Codding, G. (1991) 'The evolution of the ITU', *Telecommunications Policy* 15, 4: 271–85.

Cook, P. and Kirkpatrick, C. (eds) (1990) *Privatisation in Less Developed Countries*, Brighton: Wheatsheaf.

Cowhey, P. (1990) 'The international telecommunications regime: the political roots of regimes for high technology', *International Organisation* 44, 2: 169–99.

Curwen, P. (1995) 'Telecommunications policy in the EU: developing the information superhighway', *Journal of Common Market Studies* 33, 3: 331–60.

Dore, J. (1994) 'The European energy charter: a breakthrough in east–west energy trade?', STEEP Discussion Paper 18, Science Policy Research Unit.

Downs, G., Roche, D. and Barsoom, P. (1996) 'Is the good news about compliance good news about cooperation', *International Organisation* 50, 3: 379–406.

Drake, W. and Nicolaidis, K. (1992) 'Ideas Interests and Institutionalisation: "Trade in services" and the Uruguay round', *International Organisation* 46, 1: 37–100.

Gardner, Richard N. (1969) *Sterling–Dollar Diplomacy. The Origins and the Prospects of Our International Economic Order*, New York: McGraw-Hill.

Giddens, A. (1994) *Beyond Left and Right*, Cambridge: Polity Press.

Grubb, M. (1992) *The Earth Summit Agreements: a Guide and Assessment*, London: Earthscan.

Hannah, L. (1979) *Electricity Before Nationalisation: A Study of the Development of the Electricity Supply Industry in Britain to 1945*, London: Macmillan.

Harbutt Dawson, W. (1912) *Industrial Germany*, Glasgow: Collins.
Harrison, M. (1978) *An Economic History of Modern Spain*, Manchester: Manchester University Press.
Henderson, D. (1989) 'A new age of reform?' *Fiscal Studies* 10, 3: 72–90.
Hirst, P. and Thompson, G. (1996) *Globalisation in Question*, Oxford: Blackwell.
Hoekman, B. and Kostecki, M. (1995) *The Political Economy of the World Trading System from GATT to the WTO*, Oxford: Oxford University Press.
Holmes, P., Kempton, J. and McGowan, F. (1996) 'Telecommunications and international competition policy: lessons from the EU, prospects for the WTO', *Telecommunications Policy* 20, 10: 755–67.
Holmes, P., McGowan, F. and Lehmann, A. (1996) 'International Competition Policy', mimeo.
Hufbauer, G. and Schott, J. (1993) *NAFTA: An Assessment*, Washington (DC): IIE.
Ikenberry, G.J. (1990) 'The international spread of privatization policies: inducements learning and "policy bandwaggoning"', in E. Suleiman and J. Waterbury (eds) *Privatization and Public Sector Reform*, Boulder (CO): Westview.
Jacobson, H. (1973) 'ITU, a pot-pourri of bureaucrats and industrialists', in R. Cox and H. Jacobson (eds) *The Anatomy of Influence*, Yale: Yale University Press.
Jackson, J. (1989) *The World Trading System: Law and Policy in International Economic Relations*, Cambridge (MA): MIT Press.
Jones, R.B. (1995) *Globalisation and Interdependence in the World Economy*, London: Pinter.
Kay, J. and Vickers, J. (1988) 'Regulatory reform in Britain', *Economic Policy* 8: 285–351.
Kiplagat, B. and Werner, M. (eds) (1994) *Telecommunications and Development in Africa*, Amsterdam: IOS Press.
Krasner, S. (1983) *International Regimes*, Ithaca (NY): Cornell University Press.
—— (1985) *Structural Conflict: the Third World against Global Liberalism*, Berkeley (CA): University of California Press.
—— (1991) 'Global Communications and National Power', *World Politics* 43, 3: 336–66.
Krugman, P. (1994) 'Globalisation and the Inequality of Nations', *CEPR Discussion Paper 1015*, London: CEPR.
Kurisaki, Y. (1995) 'The changing role of telecommunications in the economy: globalisation and its impact on national telecommunications policy', *OECD Working Papers* 3, 79.
Lanthier, P. (1989) 'Multinationals and the French electrical industry, 1889–1940', in A. Teichover, M. Levy-Leboyer and H. Nussbaum (eds) *Historical Studies in International Corporate Business*, Cambridge: Cambridge University Press.
Majone, G. (1994) 'The rise of the regulatory state in Europe', *West European Politics* 17, 1: 77–101.
McCormick, P. (1994) 'Caribbean telecommunications privatisation and regulation in the information age', *Bulletin of East Caribbean Affairs* 19, 3: 17–27.
McGowan, F. (1992) 'Infrastructures and Development', in *Global Perspectives to 2010: Tasks for Science and Technology (vol. 16)* Brussels: FAST.
—— (1993) *The Struggle for Power in Europe*, London: RIIA.
McGowan, F. and Seabright, P. (1995) 'Regulation in the European Community and

its impact on the UK', in M. Bishop, J. Kay and C. Mayer (eds) *The Regulatory Challenge*, Oxford: Oxford University Press.

McGowan, F. and Wallace, H. (1997) 'Towards a European regulatory state', *Journal of European Public Policy* 13, 4: 560–77.

Moeller, J. (1993) 'US involvement in foreign utilities under the Public Utilities Holding Company Act of 1935 and the Energy Policy Act of 1992', *Law and Policy in International Business* 24, 2: 503–52.

Mosley, P. (1989) 'Privatisation policy based lending and World Bank behaviour', in Cook, P. and Kirkpatrick, C. (eds) *Privatisation in Less Developed Countries*, Brighton: Wheatsheaf.

Murphy, C. (1994) *International Organisation and Industrial Change: Global Governance since 1850*, Cambridge: Polity Press.

Nellis, J. (1988) 'Contract plans and public enterprise performance', World Bank Discussion Paper 48.

Noam, E. (1992) *Telecommunications in Europe*, New York: Oxford University Press.

OECD (1992) *Regulatory Reform, Privatisation and Competition Policy*, Paris: OECD.

Ohmae, K. (1992) *The Borderless World*, London: Fontana.

Ramamurti, R. and Vernon, R. (eds) (1991) *Privatization and Control of State Owned Enterprises*, Washington (DC): World Bank.

Sampson, A. (1973) *Sovereign State: the Secret History of ITT*, London: Hodder.

Scheinmann, L. (1973) 'IAEA, atomic condominium', in R. Cox and H. Jacobson (eds) *The Anatomy of Influence*, Yale: Yale University Press.

Sobel, R. (1982) *ITT: the Management of Opportunity*, London: Sidgwick and Jackson.

Sprinz, D. and Vaahtoranta, T. (1994) 'The interest based explanational of international environmental policy', *International Organisation* 48, 1: 77–105.

Stewart-Smith, M. (1995) 'Private financing and infrastructure provision in emerging markets', *Law and Policy in International Business* 26, 4: 987–1012.

Stoffaës, Ch. (ed.) (1995) *L'Europe à l'épreuve de l'intérêt général*, Paris: ISUPE.

Tendler, J. (1968) *Electric Power in Brazil: Entrepreneurship in the Public Sector*, Cambridge (MA): Harvard University Press.

Vuylstecke, C. (1988) 'Techniques of privatisation of state owned enterprises, vol. 1', *World Bank Technical Paper 88*.

Webster, R. (1975) *Industrial Imperialism in Italy*, Berkeley (CA): University of California Press.

Wilkins, M. (1970) *The Emergence of Multinational Enterprise: American Business Abroad from the Colonial Era to 1914*, Cambridge (MA): Harvard University Press.

—— (1974) *The Maturing of Multinational Enterprise: American Business Abroad from the 1914 to 1970*, Cambridge (MA): Harvard University Press

Williamson, J. (ed.) (1994) *The Political Economy of Policy Reform*, Washington (DC): IIE.

Wionczeck, M. (1964) 'Electric power: the uneasy partnership', in R. Vernon (ed.) *Public Policy and Private Enterprise in Mexico*, Cambridge (MA): Harvard University Press.

World Bank (1994) *World Development Report: Infrastructure for Development*, New York: Oxford University Press.

Zamagni, V. (1993) *Economic History of Italy 1860–1990*, Oxford: Clarendon Press.

9

SPLINTERING NETWORKS

The social, spatial and environmental implications of the privatization and liberalization of utilities in Britain

Simon Guy, Stephen Graham and Simon Marvin

Introduction

The past fifteen years have seen a remarkable shift in the regulation and management of infrastructure networks on the United Kingdom. Wide scale privatization and liberalization of telecommunications (1984), gas (1986), water and waste (1989), and electricity (1990–1) networks have transformed the United Kingdom into a vast laboratory, with innovative new techniques for controlling urban technical networks being implemented and evaluated (Ernst 1994). National regulatory offices have been brought in to oversee this shift towards competition in British utilities. Over a brief spell of frantic activity, new technologies, regulatory regimes, environmental priorities, spatial 'footprints', management styles, marketing strategies and commercial goals have emerged. These have dramatically reconfigured patterns of infrastructure provision and management within British cities (Marvin and Cornford 1993).

The delivery of vital utility services through regulated markets has major implications for both production and the economic development of cities, and for consumption and urban social development (Graham and Marvin 1995). Yet these implications have gone virtually unnoticed in urban studies and policy research. Almost all analyses of urban processes largely ignore the critical importance of infrastructure provision in urban economic development and the production of the built environment (Diamond and Spence 1989, Vickerman 1991). Where 'infrastructure' is treated, attention usually centres on transport and, occasionally, telecommunications (Vickerman 1991). Here, simple linear relations between the public provision of infrastructure and urban and regional development still dominate analysis (Peck 1996).

Treatment of social access to urban services, meanwhile, has been almost exclusively confined to collective welfare facilities (Pinch 1986), schools (Rogerson *et al.* 1996), health facilities (Knox 1995), retailing (Wrigley and Lowe 1996) and the broader 'consumption cleavages' relating to housing tenure (Saunders 1989). The huge resurgence of consumption debates in urban studies associated with the rise of postmodern discourses has totally ignored the consumption of utility services (see, for example, Miller 1995, Crew and Lowe 1995). Finally, public awareness of the impact of privatization and liberalization in the United Kingdom has tended to be limited to media hype around the ensuing share options, job losses and rising executive pay and company profits. Thus, there has been very little debate or understanding of the wider socio-technical ramifications of such a swift liberalization of the ownership and management of urban infrastructure networks, either in the academic or in the media arenas.

This chapter seeks to start developing just such an understanding. In particular, it aims at critically problematizing the whole approach of urban analysts to the creation and development of urban technical networks. Rather than take for granted universal access to utility services, by simply assuming infrastructure provision to be a technologically homogeneous and socially cohesive activity, we point to the emergence of key disjunctures between social, environmental and commercial priorities introduced by the privatization and liberalization process. Critically, we stress the urgent need for spatial sensitivity in unpacking the social, environmental and commercial impact of different regimes of infrastructure provision.

Focusing on electricity, telecommunications, gas, waste and water networks, we develop an analytical framework for analysing the urban effects of the radical transformation of technical networks over the last fifteen years in Britain. The chapter is structured as follows.

In the first part, we analyse the transformation of urban technical systems in the United Kingdom since the implementation of the Conservative privatization process of the 1980s. Focusing on individual sectors, we trace the process by which new 'splintered' networks are rapidly replacing the largely-standardized and homogeneous networks that developed over the postwar period.

In the second part, we build upon this analysis to identify the social, economic and environmental logics emerging from this process of rapid change. Here we highlight the heterogeneous social and environmental impact of 'splintered' urban networks by comparing and contrasting socio-economic processes of 'cherry picking' – the growing competitive focus of privatized utility companies on lucrative niche markets – and 'social dumping' – the easing out of economically marginal domestic markets – with the environmental benefits of increased network efficiency.

We conclude by briefly discussing directions for future research in the field.

The transformation of urban technical networks in the UK

The United Kingdom provides an excellent example of the emerging patterns of global-localization in urban technical networks. In this section, we aim to illustrate how new splintered networks, operating in the privatized and liberalized context, are rapidly replacing the standardized and homogeneous networks that developed over the postwar period. Utilities have always developed unevenly in the UK in institutional, spatial and temporal terms, with considerable variation between services. Nevertheless, the three-phase model summarized in Table 9.1 provides a useful framework against which to analyse the recent evolution of utility networks (see Graham and Marvin 1995).

Building networks

Localization was the initial stage of network development. A patchwork of small private and municipally-owned enterprises were responsible for the establishment of urban networks. Rapid urbanization from the 1840s was based on an increasingly dense lattice of technical support services – initially water, waste and gas, later followed by electricity, transit systems and the telephone (Ausubel and Herman 1988, Tarr 1984, Tarr and Dupuy 1988). This 'bundle' of support services facilitated initial urban growth and development, overcoming the social, environmental and economic constraints to the formation of dense urban-industrial agglomerations.

Early networks tended to follow broadly similar patterns of development. They focused on dense urban cores with the highest demands, targeted at particular types of local markets. The telephone and electricity operators looked to large business and commercial users (Forty 1986, Preston 1990); water was directed towards households in the interests of improving public health (Chant 1989); gas was initially focused on public lighting. The configuration of these early, local networks were largely shaped by local economic, social, political and spatial considerations. Consequently, there was enormous diversity in tariffs, levels of connection, the spatial extent of networks and quality of service (Dimcock 1933).

Expanding networks

Nationalization was the antithesis of this model of network development. The postwar centralization of utility networks within large publicly-owned national corporations swept away local control (Sleeman 1953). There was increasing concern that the small fragmented 'islands' of utility networks imposed serious constraints on levels of national economic and social development (Hall and Preston 1988). Variable tariffs, non-standardized

Table 9.1 Phases of urban technical development

	Localization 1840s–	*Nationalization 1940s–*	*'Global-localization' 1980s–*
Infrastructure provider	Local private and municipal companies	Large nationalized public corporations	Mix of competing private companies
Orientation of providers	Local growth large users	National economic development, supply driven, cross subsidy	Rate of return, 'cherry picking' premium markets and demand driven
Scale of networks	New networks focused on urban cores	Linking and extending local networks into regional and national grids covering urban systems	Utility patchwork replicating networks plus renewal and open access to old networks
Network technology	Heavy, new intra-urban, core	Lighter inter-urban, linking, one network, consolidation	Renewal and new, local global, focus on cities plus inter-urban and international, networks
Type of regulation	Parliamentary acts	Central government direction and internal public management	Regulated liberalized markets
Objectives of regulators	Minimize costs	Universal service at standardized tariffs and technologies	International competitiveness of national economy and competition between cities
Production dimensions	Local economic development	National economic development, equalization of space and economies of scale	Rebalancing of tariffs, recommodification, local growth coalitions, cross investment
Social dimensions	Health, safety, lighting	Universal access standard services mass domestic markets	Social polarization, fragmentation, 'pay per' and social dumping

systems, differential types of services, low levels of domestic connections and inefficient systems were all widely perceived as serious constraints on national economic growth and the comparative economic performance of Britain against its closest competitors (Dimcock 1933).

In the interwar period, this led to an uneven and highly contested shift towards greater central planning in network management. From the late 1900s, there were a series of initiatives to develop more nationally-integrated, inter-urban networks. Nationalization of the telephone network in 1911, the creation of regional electricity networks in 1926, and the creation of regional water undertakings, were all part of this process of imposing a greater degree of standardization and central coordination over network development.

During the Second World War, there was little resistance to central government intervention in the provision and management of networked services, which demonstrated the economic benefits of a more centralized approach to utility planning. The need for further rationalization of the utility industries thus gained recognition, and the strength of the postwar Labour victory finally swept alternative models aside. With the nationalization of key aspects of the British economy, large public corporations were given sole responsibility for infrastructure provision. Nationalized corporations were supposed to be free from ministerial or political influence, to efficiently, and effectively, manage public service monopolies in the national interest (Coombes 1970, Reid and Allen 1970, Sleeman 1953).

While the technological parameters of network management were ill-defined, it was routinely accepted that public utility corporations were charged with 'rolling out' national networks, extending networks into rural areas, and completing the connection of domestic households on a 'universal service' basis. Tariffs and levels of service were standardized as the utilities created networks to serve an increasingly national rather than local economic space. The extension of utility networks into the domestic sector helped to create new markets for consumer goods, underpinning national postwar industrial development, the expansion of associated consumer markets like domestic appliances, and social 'modernization'.

A new logic of network management slowly emerged during the nationalized period of utility control. Perhaps the most important dynamic guiding the evolution of these technical networks was an extremely powerful supply-oriented logic of network development (Graham and Marvin 1995). Expansion of utility networks became intimately connected with the drive to improve national economic performance and quality of life. Levels of energy consumption, connection to water and waste networks and levels of telephone ownership thus became surrogate indicators for levels of national economic performance (Sleeman 1953).

In the search for greater economies of scale, for example, the electricity industry built larger power stations and upgraded the national electricity transmission network. In the space of twenty years following the Second

World War, generative capacity multiplied seventeen-fold (Reid and Allen 1970: 9). Driven by the basic assumption that economic growth would generate new demands for utility services, network providers became locked into a logic of network management that focused on improving the quantity and quality of the supply of networked services. Major investments in national transport, energy and telecommunications services were made during this period in order to develop standardized supply systems and support regional economic equalization.

Guided by this new logic, the specific needs of individual cities had little impact on the process of network provision and management (Graham and Marvin 1994, 1995). Although cities were critically important as centres of growing demand, the management of networked services was primarily concerned with nation-wide economic development, with national priorities guiding strategic decisions. Regional arms of the utilities began to act merely as sales and marketing departments for powerful central planning agencies. With municipal control extinguished through postwar nationalization, relatively little room remained for urban politicians and planners to intervene within the decision-making process guiding infrastructure provision. As the hidden, unseen, quiet and unobtrusive networks expanded, providing utility services for customers at standardized prices and tariffs, little local interest was shown in the social, economic and environmental profile of networked services within cities.

Privatization

Privatization has radically challenged the logic of network management established during the nationalized period. These changes are having significant changes for the development of technical networks in Britain (Marvin and Graham 1994a).

During the 1970s, the nationalized period of utility development came under increasing pressure as the supply-side logic of network management became harder to sustain (Vickers and Yarrow 1989). The energy crisis of the early 1970s exposed the costs of a fuel strategy purely based on supply-side measures. Increasing constraints on public expenditure meant that the funding of supply-oriented measures became severely strained, for example, in telecommunications where the need for modernization was continuous (Houlihan 1992). There was increasing concern about the failure of supply-oriented public utilities. On the social side, there was increasing concern about the impacts of rising energy prices on the fuel poor (Boardman 1991). Environmentalists, meanwhile, were critical of the failure to seriously pursue energy and water conservation and efficiency measures, while the development of new power stations and coal fields became increasingly controversial. In its commitment to achieving economies of scale in the generation and sale of electricity, its monopolistic, prescribed pricing structures and its inflated capacity, the Central Electricity Generating Board (CEGB), in particular, came to be seen as an arrogant producer, unresponsive to local demands (Bonner 1989).

At another level, there was criticism of a perceived infrastructure crisis due to the failure to invest in renewing old and ageing energy and water networks, while telecommunications infrastructure lagged behind competitor countries (Cowie *et al.* 1984). In response, the government attempted to improve the economic efficiency of the utilities sector through financial controls, targets and increased rates of return. It was widely felt that the sector was inefficient, lacked clear objectives and operated outside wider political control (Vickers and Yarrow 1989). Although various measures were considered for restructuring the utilities sector through the 1970s, there was widespread failure to implement comprehensive operational changes.

During the 1980s, a solution to the problems of the utility sector developed around the Thatcherite privatization project. The basic rationale for privatization was based on a critique of the nationalized sector. This comprised a number of key justifications: the sector was seen as inefficient and overstaffed, unaccountable, union-dominated and technologically stagnant. In the early years, the debate was simply based on a shift from public to privatized monopolies (with the initial privatization of British Gas and British Telecom). The central assumption was that privatized management would transform the nature of the industries, so improving levels of service, increasing efficiency and producing more responsive, competitive industries. The new industries operated within a regulated framework designed to protect consumer interests and promote the efficiency of the sector. But increasing criticism of the performance of privatized monopolies created the pressure, to increase levels of competition, to allow domestic mergers and takeovers and to relax restrictions on foreign capital from making inroads into the UK.

Splintering networks

Although privatization and liberalization has developed unevenly across the utilities sector, we can explore how the two processes have increasingly led to the 'splintering' of utility networks. The complex transformations underway are best summarized diagrammatically, in Figures 9.1 and 9.2, which provide a schematic representation of transformations in the electricity and telecommunications sectors respectively.

The inherent transportation difficulties involved mean that competitive pressures have been hardest to introduce in to the water sector. Although OFWAT (the water industry's regulatory office) monitors the comparative economic efficiency and performance of the water companies, the largely regional structure of the water networks and the difficulties of allowing competitors access to another company's infrastructure, have seriously constrained competition. However, OFWAT has encouraged competition on the boundaries of water companies' service areas. A large user siting a development on the boundary between two companies could choose their supplier, while another alternative could include a supply from the British Waterways Board through the canal network.

Competitive pressures have been more successfully introduced into the energy sector. Although British Gas was privatized as a vertically-integrated company, regulatory pressures have forced the company to take measures to allow competitors access to the transmission and distribution network. Increasing controversy about the charges for access to the incumbent network have created demands for a more transparent pricing structure – British Gas has been forced to split its transmission structure into a separate business. Competitors now supply over 50 per cent of the industrial and commercial market. In 1998, limits on competition were reduced further when the introduction of competitive supply within the domestic sector further eroded the British Gas market.

These issues are easier to resolve in the electricity sector, where energy is inherently more transportable. Figure 9.1 illustrates how, at privatization, the

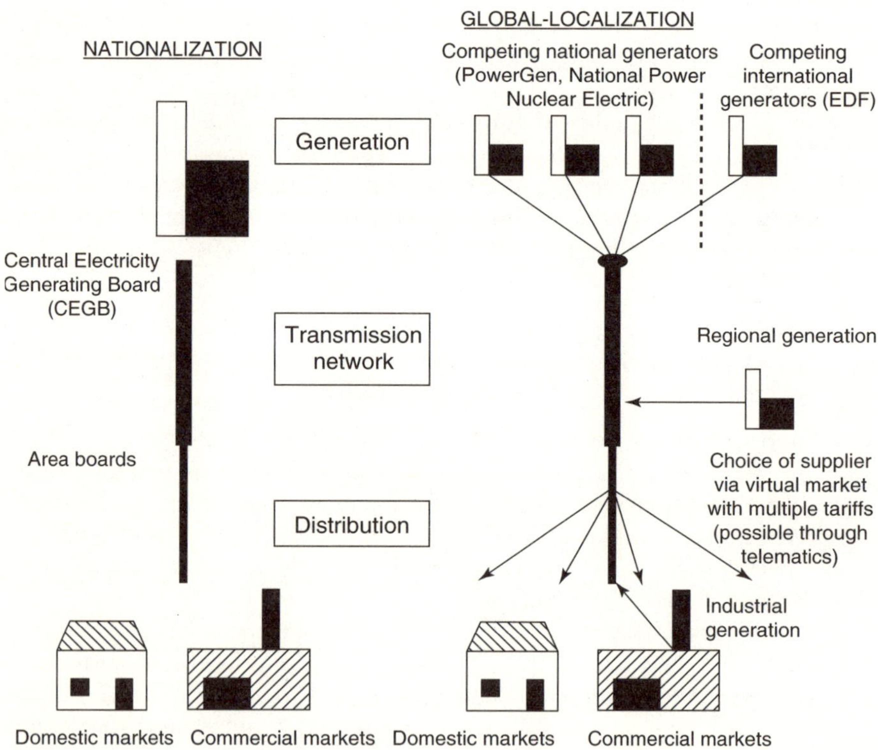

Figure 9.1 The transformation of the electricity network

electricity sector was broken down into different elements (generation, transmission and distribution). After the criticism of the constraints the privatized British Gas was able to use to delay competitive pressure, there was a conscious attempt to introduced more competition into the electricity sector. In particular, new technologies have played a crucial role in fostering competition in electricity supply. Large users are able to choose their supplier utilizing 'smart metering' technologies to monitor consumption and send readings electronically to their supplier. A new 'virtual market' in electricity supply has been created by overlaying a new telecommunications infrastructure over the electricity distribution network. Real time flows of information between 'smart cards' and central computers allow the flows of electricity to be monitored and recorded, allowing complex billing, tariffs and competition to occur without the need for separate electricity networks and connections to customers. Finally, the domestic market is being opened up for competition through regulation in 1998.

The shift towards competition and new entrants has been most pronounced in the telecommunications sector. Figure 9.2 demonstrates the extremely

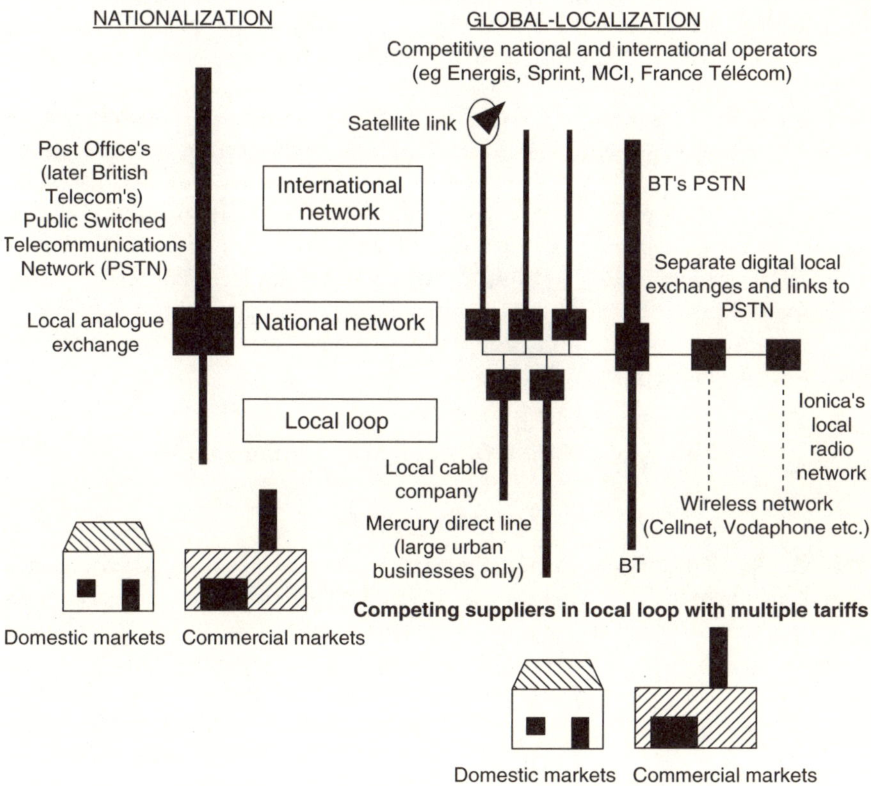

Figure 9.2 The transformation of the telecommunications network

complex patterns of network splintering here. Although the 150 or more new entrants into the telecommunications market can interconnect with the local and trunk parts of BT's original Public Switched Telecommunications Network (PSTN), they have also started to develop their own competitive trunk and local networks. At the local level, cable TV companies offer telephony as well as TV services, a company called Ionica plans to offer local phone services by fixed radio links and several mobile systems compete for market shares. The largest mainstream competitor, Mercury, has also directly cabled large users in city centres. Regionally, the cable TV companies are offering services through linkage between franchise areas while the regional electricity companies have now started offering services in their franchise areas.

Nationally, Mercury have a trunk network layered over the rail network. Energis, a telecom company owned by NGC (the national electricity grid company), operate a system layered over the electricity supergrid. British Waterways and British Rail are also entering telecommunications markets, using their own infrastructures as the basis for laying telecommunications cables. Finally, at the international level, a wide range of companies, such as Sprint, MCI, AT&T, and France Télécom, are offering competitive telecommunications services, focusing especially on large London-based businesses.

Thus, utility networks are being splintered in different ways in each sector. Basically in water, gas and electricity markets new entrants are being allowed access to existing networks. New technology has played a central role in reducing the transactions costs involved in competitive utility markets. Technologies like electronic data interchange (EDI) and 'smart meters' enable competitive supplies to pass through monopolistic systems overlain with telematics networks. Telecommunications has a completely different logic, with open access to BT networks, the development of competing telecommunications infrastructures and a whole range of new entrants at local, national and international levels.

Emerging logics of network management

The logic of profit

The whirlpool of change initiated by the introduction of privatization and liberalization of utility markets has radically reoriented the priorities and practices of utility companies. The ensuing splintering of urban technical networks is, in turn, having profound impacts on infrastructure provision in British cities. Assessing the social, economic and environmental benefits and disbenefits of this process is not as straightforward as many commentators on the privatization of public utilities propound. While one camp decries the inequities introduced by privatization (Ernst 1994), another celebrates the fresh opportunities for the consumer and shareholder (Saunders and Harris

1994). But such totalizing verdicts homogenize a complex and variegated set of processes. Closer investigation of the emergence of new styles of utility network management reveals complex patterns of interwoven social, economic and environmental change.

The key shift instigated by the privatization and liberalization of utility markets has been the replacement of the ethic of public service – the ideal of affordable, reliable, universal access to utility services at constant tariffs for all, irrespective of income or location. Replacing this has been the goal of profitability. The overriding aim of British utility companies today is the maximization of profits for shareholders and (increasingly global) financial investors. The result is mounting pressure on utility companies to find novel ways of extracting surplus value from their networks (or users), while reducing their dependence on the vagaries of the fortunes of the local economies within which their networks are, quite literally, 'embedded' (Cox and Mair 1988).

This quest has led to the emergence of three related social, economic and environmental logics which, we argue, increasingly guide the management and development of urban technical networks in the United Kingdom. These are 'cherry picking', 'social dumping' and 'demand-side management'. Table 9.2 illustrates the characteristics of these three logics. These new logics of network management are, in turn, beginning to reveal themselves in urban areas, as privatized utilities adjust to the new uncertainties created by regulated markets for utility services.

Cherry picking

The privatization and liberalization of utility marketplaces have encouraged a keen sensitivity to the market potential and profitability of infrastructure provision. The provision of electricity, gas, water, telecommunications is no longer seen as a functional utility service. Instead, as part of the shift toward 'business process re-engineering', private utility companies now see themselves as selling quality services (Goss 1995). Using geodemographic analysis and consumer profiling techniques, these services are now carefully targeted through 'lifestyle marketing' at profiled socio-economic groups and places (Longley and Clarke 1995, Martin 1996). 'By general consensus', writes Alicia Clegg (1995: 17), 'the domestic customers who'll benefit from market liberalisation will be [the privileged socioeconomic groups] – and it is these privileged groups that we can expect the new entrants to target with price packages, innovative services and customer loyalty bonuses'.

Regional Electricity Companies, for example, no longer simply see themselves as supplying standardized kilowatt hours; they use the transactional capabilities of information technologies to sell a disaggregated range of 'energy services' – the ability to boil kettles, heat bath water, light offices and factories, etc. (Owen 1994). As Small (1996) recently argued, domestic

Table 9.2 New logics of network management

	Feature		
	'Cherry picking'	*'Social dumping'*	*'Demand-side management'*
Objective	Increase profitability by concentration on most lucrative customers	Ease out of service unprofitable and marginal customers so improving efficiency and profitability	Encourage consumers to modify consumption of water, electricity and gas to increase profitability
Spatial	Hot spots: e.g. the City of London	Ghettos: e.g. inner cities, rural areas	Targeted on over and under stretched elements of networks
Customers	Large commercial users	Poor, elderly, etc.	Large customers
Nature of service	Choice and competition	Trapped, disenfranchised	Packaged services
Service measures	Cost, quality	Prepayment, disconnection	Comfort and efficiency
Costs	Downwards	Increases	Shared savings
Impacts	More competition in hot spots	Societal costs of 'switching out' people from their means of social and economic participation, and public health risk	Reduced supply-side infrastructure investment, brand loyalty, reduced consumption

competition in electricity in 1998 will allow electricity companies to target affluent, profitable markets, in the same way that telecom companies currently do with premium services like mobile and discount schemes to support customer loyalty:

> Encouraging new energy demand means catering for demands with special applications – be it orchid houses, swimming pools, or applications yet unknown. Wireless technology [for monitoring use] is well-suited to reaching a small number of households at low marginal cost.
>
> (Small 1996: 21)

Thus, like any other commercial business, privatized utilities are now keen to tailor their products and services to the most lucrative sections of the market, to deal with different customers in very different ways. Customers who consume regularly (and healthily), pay reliably and utilize direct-debit facilities through bank accounts, are now the focus of concern. No longer classifying 'utility users' simply in terms of their technical profile – their rate of consumption – utilities now use very sophisticated 'geodemographic' profiles assessing their commercial value, lifestyle and value-added potential. These are carefully mapped using the types of geographical information systems (GISs) and geodemographic consumer profiles (like ACORN and MOSAIC) traditionally used in the retailing industry (Batey and Brown 1995).

The operational goals of utility companies are, then, to use these new analytical techniques to attract, retain and satisfy these key customers through quality utility services in an increasingly competitive marketplace. Sophisticated customer information systems, linked into call distribution systems, allow telephone enquiries centres – which are increasingly located away from the 'utility patch' – to maintain this disaggregated treatment of customers. A new system at South West Water, for example, means that

> when a customer rings, just by giving their name and postcode to the member of staff allows all account details (including records of past telephone calls, billing dates and payments, even scanned images of letters) to be displayed. This amount of information enables staff to deal with different customers in different ways. A customer who repeatedly defaults with payment will be treated completely differently from one who has only defaulted once.
>
> (*Utility Week* 1995)

In the telecoms arena, a cable company might use geodemographics to target its mainstream TV services in one area it categorizes as 'Working Hard' ('blue collar, low education, mostly female head'), whilst it targets its new Internet services to leafy suburbs marked 'Diamonds-to-Go' ('two or more adults 1 or more kid, highly affluent') (Birkin 1995). When BT lose a phone subscriber to a cable company they use telemarketing to target only those who spend on average over £60 per quarter, being glad to lose lower spenders (Graham and Marvin 1996).

In the electricity sector, meanwhile, the new marketing strategies targeting affluent customers include free energy audits, 'smart home' application like remotely-controlled dishwashers, individually-tailored tariffs (reduced through direct debit billing), swift personal advice and trouble-shooting. Regional Electricity Companies are already installing optic fibre alongside their lines in affluent areas which, beyond their use for remote monitoring of consumption, will allow them to deliver telecoms services (Small 1996).

For example, London Electricity is now offering a data management service which will allow selected customers to receive consumption and tariff information not only on electricity but also on gas and water, via EDI, linking the customers' computerized meter with the company's computer system (Woodford 1996). Administered by individual account managers, London Electricity are seeking to build 'strong business partnerships (by) supplying increasingly sophisticated added-value services' (Steele 1996: 13). As Carl Weinberg puts it, utilities 'have learned that their survival in a competitive world depends on an ability to understand what it is their customers want', and that 'customers are not necessarily interested in low-cost kilowatt hours, but instead in low-cost, high-quality energy services' (Weinberg 1994: 291).

Careful spatial targeting is inextricably woven in with this social targeting. Private utility companies are keen to focus their investment in geographically bound areas, or 'hot spots', where potential profits are maximized (through high demand with low costs). The geodemographic profiling above inevitably leads to utilities focusing investment, service innovations and marketing into the more affluent areas of cities. GISs, in particular, are being used to plan network extension across all the utilities, funnelling investment into the 'profit rich' spatial areas of cities while neglecting the rest. On the economic front, large, profitable industrial users and companies become the focus of the cherry picking process. This creation of 'customised spaces' (Peck 1996) linked to advanced infrastructure networks applies to existing and inward investing firms, as is demonstrated by the extremely specialized and tailored treatment received from utilities by recent inward investors in the North East, like Siemens, Fujitsu and Samsung. Another excellent example here is the City of London where BT, Mercury, cable companies and US operators such as Colt and MFS have built specialized, high capacity, optic fibre networks. Along with many other 'resellers' of telecom capacity, they now fight for the highly lucrative and specialized custom of multi-national financial services companies, with their extremely high demands for global telecommunications services (Graham and Marvin 1994).

This process of cherry picking profitable social areas and industrial users in turn reflects the growing internationalization of the utility marketplace, which is evolving in parallel with global shifts toward privatization, liberalization and international flows of capital, expertise and technology. Keen to seek out ever-more lucrative markets, national utilities are reaching out beyond national boundaries in an effort to forge global alliances, acquisitions and mergers – a basis for the elaboration of transnational networks of power and communications. Similarly, energy and water companies are taking on an increasingly international profile, with British companies active in the developing markets of Eastern Europe and the Far East, while other European companies take-up the commercial opportunities offered by the liberalized British market. Within the UK, since privatization, a continuous and rapid process of takeover and investment activity has occurred, opening up urban infrastructure systems as commodities to be traded and linked into global flows of capital, expertise and technology. For example, UK cable has been developed almost entirely by US cable and telecom companies; North West Water has taken over the Norweb electricity company (forming 'United Utilities' – a single water and energy company); the French municipal giant Compagnie Générale des Eaux has taken over Northumbrian Water; and the US Southern Electric company recently acquired South West Electric.

Social dumping

The corollary of a cherry picking strategy is the 'social dumping' of unprofitable consumers and the withdrawal from zones of little commercial opportunity (Graham and Marvin 1994). In rural areas and disadvantaged inner cities, competition remains scarce, with poor levels of prospective income and high operational costs deterring new investment (Marvin 1994). At the same time, the gradual removal of the cross-subsidies that underpinned the 'rolling out' of national utility networks in the postwar period highlights the 'real' cost of servicing commercial 'cold spots'. This is forcing up prices and lowering the quality of local infrastructure and services through a process known as 'tariff rebalancing' (bringing charges into line with costs). With utility companies globalizing their operations in search of higher returns, the capacity of disadvantaged communities to maintain competitive and affordable utility infrastructures and services is diminishing. Inevitably, this exacerbates wider trends towards socio-spatial polarization within cities (Knox 1995, Rogerson *et al.* 1996).

This process graphically highlights the changing space-economy that results from the splintering of infrastructure networks. As marginalized urban and rural zones become disconnected or 'dumped' from globalizing networks of power and communications, whole regions are becoming more socially and economically polarized in terms of access to, and use of, utilities (National Consumer Council 1993).

The poorest customers are likely to be low consumers of utility services with poor payment records (see Figure 9.3). As such, they are viewed as an impediment to the global aspirations of utility companies, who are seeking to minimize technical and administrative expenditure, while maximizing income. While British Gas estimate that they spend £10 per year in bill management for customers on direct debit, customers who go through all the procedures short of disconnection are estimated to cost the company £100 per year (Robertson 1995). With universal service obligations being gradually eroded, utility companies are striving to cut these costs, either by levering poor domestic customers off their networks, or by installing prepayment meters (which avoid the costs of bad payers, by allowing real time payment for utilities, via 'smart cards' that are 'topped up' through cash payments at post offices).

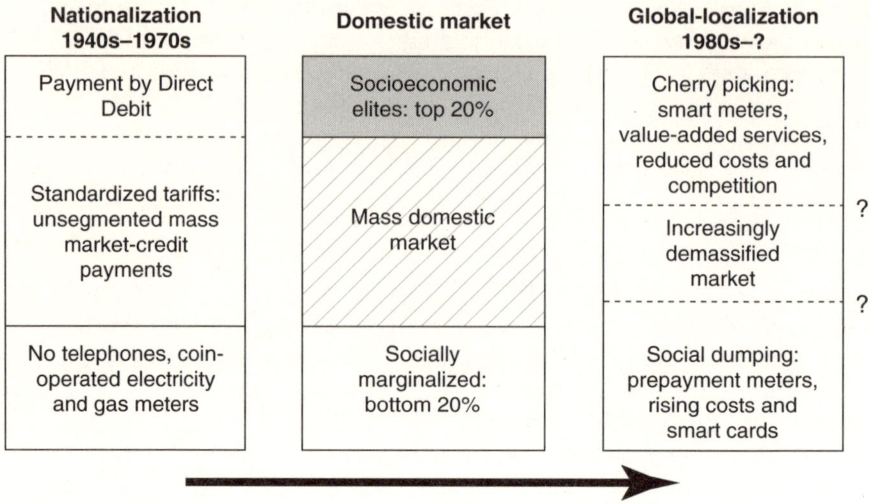

Figure 9.3 The splintering of domestic utility markets associated with privatization/liberalization

Recent tariff reforms have reduced cross-subsidies, so supporting this socio-spatial polarization. Service charges have risen at rates higher than the rate of inflation across all utility sectors. These rises have had a differential impact on different classes of user (Marvin 1994). A recent survey of gas consumers in the competitive trial in the South West of England, for example, is revealing. It showed that an affluent gas consumer, subject to competitive cherry picking and able to pay by direct debit, could obtain competitive gas supplied from CalorTex for 36 pence per therm. A marginal consumer, on the other hand, unable to tempt in new competitors, would be trapped as a customer of British Gas, forced to use a prepayment meter, and be charged 81 pence per therm (Durham and Barnett 1996).

Such processes can be seen right across the utility sectors. Since privatization, BT has been allowed to radically increase both line rental charges and local call charges well above the rate of inflation, whilst reducing national and international call costs. Local call charges rose by 80 per cent in real terms between 1985 and 1993, while international calls charges declined by 20 per cent (Ernst 1994). This hits poorer, local users hardest. BT also require high deposits from customers with poor payment records. The growing use of prepayment meters, in particular, not only supports higher charges for the poorest consumers, but also serves to further marginalize the utility poor by masking levels of disconnection. Inability to pay for energy, water or communications does not now require action on the part of the utility company, as users effectively disconnect themselves from the network in a silent, hidden process. Between 1988 and 1991, for example, gas disconnections fell from

0.4 to 0.1 per 1,000 in the UK whilst installed prepayment meters rose from 0.09 per 1,000 to 0.5 per 1,000 (Ernst 1994).

Demand-side management

Focusing exclusively on the socio-economic implications of the splintering process can, however, result in an overly negative view of the impacts of privatization and liberalization of utility networks. While worries over the fate of marginalized, 'uneconomic' spaces and users are clearly legitimate, they can mask the more positive, environmentally-beneficial effects of a more competitive approach to infrastructure provision (especially in energy and water networks). But this is not to say that privatizing utility markets leads automatically to environmental sensitivity. Rather, the environmental outcome of the splintering of networks varies according to the physical, commercial and regulatory constraints shaping different utility sectors, and to the specific operational priorities shaping local, regional and national utility strategies.

The electricity sector provides some clear examples of the environmental opportunities provided by privatization and liberalization (Guy 1994). During the nationalized era, electricity demand profiling was the responsibility of the Central Electricity Generating Board who placed more emphasis upon the avoidance of spectacular 'blackouts' than upon the less dramatic concern of distribution losses (Berrie 1992: xx). With investment in new infrastructure now in the hands of the regional electricity companies (RECs), closer attention is being paid to the operational efficiency of the distribution network feeding electricity supplies. While flat, predictable demand has always been the ideal goal of electricity planning (Nye 1992), RECs have a particular interest in minimizing distribution losses in order to avoid any unnecessary purchase of electricity from the national electricity 'pool'. This is stimulating refined management of regional supply networks. Greater efforts are being made to promote smooth local demand profiles, thereby minimizing losses of 'electrons' (and profits). At the same time the energy audits, more detailed demand profiles and multiple tariffs increasingly offered by regional electricity companies in order to win new customers, are encouraging users to alter their patterns of demand, their 'load profile', to achieve significant electricity savings (Bennell 1994).

Similarly, new tariff structures are allowing a continuously-varying price (per kWh), matching as near as possible the actual costs of generation, transmission and distribution. As Tom Berrie points out, such 'spot' pricing has benefits for producers and consumers alike, encouraging operating efficiency improvements, 'capital investment reductions, improved consumer options on supply quality or reliability and lower electricity prices' (Berrie 1992: xxvi). Dynamic pricing presents real incentives for users to tailor their electricity needs to a changing structure of supply. On over-stretched networks, the avoidance of electricity use at peak periods may help to smooth the demand-profiles, reducing the need for environmentally damaging infrastructure investment. Such a strategy was exploited by Manweb as the branch of their distribution network serving the island of Anglesey started to become 'hot'. With only two small (33 KW) sub-stations meeting a peak demand (9 MW), growing at 2 per cent per year, expensive network reinforcement, consisting of a new transformer and cables costing roughly £1 million, seemed inevitable. However, by reducing demand peaks by 1 MW, through the provision of subsidized energy-saving light bulbs, roof insulation and comprehensive draught-proofing and free energy audits, this infrastructure investment was avoided (Guy 1994).

Private water companies are, similarly, striving to improve the efficiency of their ageing networks in which leakage levels can reach 30 per cent. With OFWAT now examining efficiency in terms of water delivered to customers, rather than water supplied into the system, a new emphasis on demand management and planning has surfaced. Hitherto, local incidences of under supply would have been solved within a wider supply strategy – a new reservoir or abstraction point. Now, with all major capital expenditure under review, such 'water stress' is being dealt with at a local level. Urban zones suffering high water stress are targeted and demand management programmes instigated through 'beyond the meter' water management in commercial and industrial premises and through the metering of domestic consumption.

Conclusions

In the new infrastructure age of global-localization, the development and management of utility networks is emerging from the shadows of the urban fabric. The vast, hidden lattices of ducts, wires and cables, long taken-for-granted as invisible, ubiquitous and rather uninteresting networks underpinning modern industrial cities, are gradually slipping back onto urban research and policy agendas.

The social, economic and environmental issues raised by the emergence of the new logics guiding infrastructure provision mean that the study of large technical systems can no longer be dismissed as a dull technical exercise. It is no longer about tracing the apparently inevitable extension of technologically standardized and spatially homogeneous utility networks. The introduction of competition between utility providers, leading to new strategies of niche marketing, social dumping and increased socio-spatial polarization, has thrown the political ramifications of contemporary utility strategies into sharp relief.

Three lessons for the study of large technical systems are critical and clear. First, changes in the provision of utilities need to be incorporated into debates about urban restructuring and changes in consumption and service provision within cities. In the current era of global-localization, there is a need to sharpen the focus of research into the changing logics guiding infrastructure provision, through increased historical, spatial and sectoral sensitivity to utility network development. Second, analysis needs to further explore the apparent disconnection of local, national and international policy-making mechanisms from these liberalized local-global processes.

Finally, urban research needs to engage in normative policy debates through which new regulatory frameworks can be developed. There is indeed a clear need for increased social and environmental regulation of utility industries in order to safeguard the interests of the utility poor and of peripheral and declining areas, while maximizing the wider environmental benefits to be captured from new commercial concerns with network efficiency.

References

Ausubel, J.H. and Herman, R. (eds) (1988) *Cities and their Vital Systems: Infrastructure Past, Present, and Future*, Washington (DC): National Academy Press.

Batey, P. and Brown, P. (1995) 'From human ecology to customer targeting: the evolution of geodemographics', in P. Longley and G. Clarke (eds) *GIS for Business and Service Planning*, Cambridge: Geoinformation International.

Bennell, P. (1994) 'Take the plunge into the 100kW market', *Energy in Buildings and Industry* (July/August): 30.

Berrie, T. (1992) *Electricity Economics and Planning*, London: Institute of Electrical Engineers.

Birkin, M. (1995) 'Customer targeting, geodemographics and lifestyle approaches', in P. Longley and G. Clarke (eds) *GIS for Business and Service Planning*, Cambridge: Geoinformation International.

Boardman, B. (1991) *Fuel Poverty: From Cold Homes to Affordable Warmth*, London: Belhaven.

Bonner, F.E. (1989) 'The electricity supply industry', *Energy Policy* 17, 1: 15–21.

Chant, C. (ed.) (1989) *Science, Technology and Everyday Life 1870–1950*, London: Routledge.

Clegg, A. (1995) 'To boldly bill', *Utility Week* (10 November): 16–17.

Coombes, D. (1971) *State Enterprise: Business or Politics?*, London: George Allen and Unwin.

Cowie, H, Harlow, C. and Emerson, R. (1984) *Rebuilding the Infrastructure: The Needs of English Cities and Towns* (report 636), London: Policy Studies Institute.

Cox, K. and Mair, A. (1988) 'Locality and community in the politics of local economic development', *Annals of the Association of American Geographers* 78, 2: 307–25.

Crewe, I. and Lowe, M. (1995) 'Gap on the map? Towards a geography of consumption and identity', *Environment and Planning A* 27: 1877–96.

Diamond, D. and Spence, N. (1989) *Infrastructure and Industrial Costs in British Industry* (Report to the Department of Trade and Industry), London: HMSO.

Dimcock, M. (1933) *British Public Utilities and National Development*, London: George Allen and Unwin.

Durham, M. and Barnett, A. (1996) 'Fuel twice the price for those on half the income', *Observer* (10 March): 8.

Ernst, J. (1994) *Whose Utility? – The Social Impact of Public Utility Privatization and Regulation in Britain*, Milton Keynes: Open University Press.

Forty, A. (1986) *Objects of Desire: Design and Society 1750–1980*, London: Thames and Hudson.

Goss, J. (1995) 'Marketing the new marketing: the strategic discourse of geodemographic information systems', in J. Pickles (ed.) *Ground Truth: The Social Implications of Geographic Information Systems*, London: Guildford.

Graham, S. and Marvin, S.J. (1994) 'Cherry picking and social dumping: British utilities in the 1990s', *Utilities Policy* 4, 2: 113–19.

—— (1995) 'More than ducts and wires: post fordism, cities and utility networks', in P. Healey, S. Cameron, S. Davoudi, S. Graham and A. Madani Pour (eds) *Managing Cities: The New Urban Context*, London: Wiley, London.

—— (1996) *Telecommunications and the City: Electronic Spaces, Urban Places*, London: Routledge.

Guy, S. (1994) 'The new energy managers? Regional electricity companies and the "logic" of demand-side management', Working Paper 46, Department of Town and Country Planning, University of Newcastle.

Hall, P. and Preston, P. (1988) *The Carrier Wave: New Information Technology and the Geography of Innovation 1846–2003*, London: Unwin.

Houlihan, B. (ed.) (1992) *The Challenge of Public Works Management: A Comparative Study of North America, Japan and Europe*, Brussels: IIAS.

Knox, P. (1995) *Urban Social Geography* (3rd edition), London: Longman.

Kraemer, K.L. and King, J.L. (1982) 'Telecommunications/transportation substitution and energy conservation, part 2', *Telecommunications Policy* (June): 87–99.

Longley, P. and Clarke, G. (1995) *GIS for Business and Service Planning*, Cambridge: Geoinformation International.

Martin, D. (1996) *Geographic Information Systems: Socioeconomic Applications*, London: Routledge.

Marvin, S.J. (1992) 'Urban policy and infrastructure networks', *Local Economy* 7, 3: 225–48.

—— (1994) 'Accessibility to utility networks: local policy issues', *Local Government Studies* 20, 3: 437–57.

Marvin, S.J. and Cornford, J. (1993) 'Regionalised infrastructure networks: policy implications', *Regional Studies* 27, 2: 159–65.

Marvin, S.J. and Graham, S. (1994a) 'Privatization of utilities: the implications for cities in the United Kingdom', *Journal of Urban Technology* 2, 1: 47–66.

—— (1994b) 'Meters and utility dualisation', *Energy Utilities* (July 1994): 22–6.

Miller, D. (ed.) (1995) *Acknowledging Consumption: A Review of New Studies*, London: Routledge.

Murdock, G. and Golding, P. (1989) 'Information poverty and political inequality: citizenship in the age of privatized communications', *Journal of Communication* 39, 3: 180–95.

Mayntz, R. and Hughes, T. (eds) (1988) *The Development of Large Technical Systems*, Frankfurt: Campus.

National Consumer Council (1993) *Paying The Price*, London: HMSO.

Nye, D.E. (1992) *Electrifying America – Social Meanings of a New Technology*, Cambridge (MA): MIT Press.

Owen, G. (1994) *From Energy Supply to Energy Services*, London: Energy Savings Trust.

Peck, F. (1996) 'Regional development and the production of space: the role of infrastructure in the attraction of inward investment', *Environment and Planning A* 28: 327–39.

Pinch, S. (1986) *Cities and Services: The Geography of Collective Consumption*, London: Routledge.

Preston, P. (1990) 'History lesson 2: some themes in the history of technology systems and networks', (PICT Paper), 1–2 March.

Reid, A. and Allen, K. (1970) *Nationalised Industries*, London: Penguin.

Roberston, F. (1995) 'Selling out to Sid', *Utility Week* (18 August): 16–17.

Rogerson, R., Findlay, A., Paddison, R. and Morris, A. (1996) 'Class, consumption and quality of life', *Progress in Planning* (January).

Saunders, P. (1989) 'Beyond housing classes: the sociological significance of private property rights in means of consumption', in L. McDowell, P. Sarre and C. Hamnett (eds) *Divided Nation*, London: Hodder and Stoughton.

Saunders, P. and Harris, C. (1994) *Privatization and Popular Capitalism*, Milton Keynes: Open University Press.

Sleeman, J. (1953) *British Public Utilities*, London: Isaac Pitman.

Small, S. (1996) 'Learning from phone wars', *Utility Week* (15 March): 20–1.

Steele, G (1996) 'Triple action', *Energy in Buildings and Industry* (February): 12–13.

Tarr, J.A. (1984) 'The evolution of urban infrastructure in the nineteenth and twentieth centuries', in R. Hanson (ed.) *Perspectives on Urban Infrastructure*, Washington (DC): National Academy Press.

Tarr, J.A. and Dupuy, G. (eds) (1988) *Technology and the Rise of the Networked City in Europe and America*, Philadelphia: Temple University Press.

Utility Week (1995) Special issue, 'IT in utilities' (19 November).

Vickerman, R.W. (ed.) (1991) *Infrastructure and Regional Development*, London: Pion.

Vickers, J. and Yarrow, G. (1989) *Privatisation: An Economic Analysis*, Cambridge (MA): MIT Press.

Weinberg, C.J. (1994) 'The restructuring of the electric utility: technology forces, R and D and sustainability', in N. Steen *Sustainable Development and the Energy Industries*, London: Earthscan.

Woodford, J. (1996) 'Electronic lines of business', *Utility Week* 22 (13–15 March).

Wrigley, N. and Lowe, M. (eds) (1996) *Retailing, Consumption and Capital*, London: Longman.

Part 3

ON THE GOVERNABILITY OF COMPLEX TECHNOLOGIES

10

THEORIES OF TECHNOLOGY AND THE ABOLITION OF NUCLEAR WEAPONS

Donald MacKenzie

In memory of Agatha C. Hughes

If theories are to be more than ways of speaking, they must help us understand concrete situations. In this chapter, I invite the reader to take part in a thought-experiment designed to examine this dimension of theoretical perspectives in the social study of technology.[1] Imagine that a prime minister or president is examining the feasibility of the permanent, world-wide abolition of nuclear weapons. She commissions you to tell her what the social study of technology has to say about the feasibility of the abolition of nuclear weapons. 'You talk much, using much jargon, about the emergence of new technologies,' she says. 'Can you not tell us, simply and clearly, what your field has to say about getting rid of them?'

I propose this as an exercise in thinking through the practical meaning of the theoretical perspectives with which our field abounds. I sketch my own answer below, focusing on four important sets of perspectives:

1 Technological systems theory, as developed above all by Thomas P. Hughes.
2 Cognitive perspectives, especially notions of tacit, local knowledge, as developed, above all, by Harry Collins.
3 'Structural sociology', in which broad category I include Marxism, modernization theory and feminism.
4 Actor-network theory, as developed by authors such as Michel Callon, Bruno Latour, and John Law.

The need for brevity means that I will be oversimplifying these perspectives, and ignoring other approaches that could be drawn on: for example, 'cultural studies' perspectives; anthropological perspectives; Bijker and Pinch's 'relevant social groups' approach; and perspectives inspired by ethnomethodology or discourse analysis. I ask only this of those who are appalled by my theoretical crudity: contribute to the discussion by showing how a more faithful, or a more diverse, theoretical analysis adds to our understanding of the feasibility of nuclear disarmament.

The *desirability* of abolishing nuclear weapons is my focus in only one, restricted sense. I do not attempt to weigh the risks of abolition (might the probability of conventional war increase? might states be more likely to turn to chemical or biological weapons?) against the risks of their indefinite continued presence (nuclear weapons accidents? theft of nuclear weapons? nuclear war through miscalculation or irrationality? and so on). Instead, my concern is with what in most discussions of comprehensive nuclear disarmament is taken as the dominant risk of the endeavour: that an abolition agreement is either circumvented by a state hiding away some weapons or is subsequently broken by nuclear rearmament.

I make a number of assumptions about nuclear disarmament. First, I assume it would be implemented gradually, perhaps over a period of twenty to thirty years (see, e.g., MccGwire 1994); it would, for example, be over that kind of period that a weapon would have to be hidden away. Second, I assume that any plausible nuclear disarmament agreement would involve more than just the dismantling of existing arsenals. I assume it would include acceptance of intrusive inspection and monitoring of suspected nuclear facilities, at least as intrusive as the mechanisms in place under current arms limitation treaties and the Nuclear Non-Proliferation Treaty. Third, I assume a prohibition on possessing or testing long-range ballistic missiles (the most important 'dedicated' nuclear weapons delivery systems), and possibly a ban on long range cruise (air-breathing) missiles as well. Fourth, I assume the existence of a well-resourced intelligence effort, including use of both human intelligence and remote surveillance, to search for violations such as the construction of unmonitored nuclear facilities. Finally, I assume that 'breakout' from a treaty to eliminate nuclear weapons would have to be covert. In other words, I assume that if attempts at violation were detected, there would be both the capacity and the willingness to take effective, limited military action against them (akin to the 1981 Israeli air-strike that rendered inoperable the nuclear reactor under construction at Osirak in Iraq).

Nuclear weapons from a 'systems' perspective

Modern technical artefacts are typically parts of technological systems. Systems, Tom Hughes asserts, manifest growth patterns driven above all by the need to maximize load factor, the ratio of average usage (which determines revenue) to peak usage (which determines the necessary level of capital investment). Technical innovation focuses around the elimination of the obstacles to growth, 'reverse salients' as Hughes calls them (see, above all, Hughes 1983).

The systems perspective implies that technologies cannot be understood in isolation, but only in their contexts, especially their systemic contexts. The outside world, for example, might have detected Saddam Hussein's nuclear weapons programme earlier had this been kept in mind. What helped keep the Iraqi programme secret was that Iraq adopted an apparently outdated method of uranium enrichment: electro-magnetic isotope separation.[2] This places enormous demands upon electricity supply: the uranium separation plant at Oak Ridge, Tennessee, consumed more electricity in 1945 than the whole of Canada produced in the Second World War (Albright and Hibbs 1991: 19). Both the International Atomic Energy Agency (responsible for Non-Proliferation Treaty inspections) and US intelligence seem implicitly to have taken for granted that no one would adopt this method of separation, so its main elements were no longer classified, there were no controls over its component technologies, and no one seems to have monitored purchases of items of these technologies (Davis and Kay 1992). Viewed in isolation, then, electro-magnetic separation was indeed an inferior and superseded technology, but viewed in the Iraqi context, in which energy was plentiful but secrecy a priority, it was an eminently sensible choice.

Nuclear weapons as the products of technological systems

If artefacts are parts of a technological system, then they can often be rendered useless by disconnection from that system: a washing machine, say, does not work if unplugged from its electricity supply. If nuclear weapons are products of, and parts of, large technological systems, could they be rendered powerless by being unplugged from those systems?

To make even a single nuclear weapon requires quantities of the order of a kilogram or more of materials that either exist in nature only in minuscule amounts (plutonium) or are more common but intermingled with other materials (uranium-235, which has to be separated from the more common isotope, uranium-238). Although state-of-the-art plutonium production and uranium separation are no longer the massive industrial activities they were at the time of the Manhattan project, the capacity to detect them is at the core of the Nuclear Non-Proliferation Treaty.

Unfortunately, control over the production of plutonium and enrichment of uranium, though clearly important, would on its own be a shaky underpinning of comprehensive nuclear disarmament. The chief problem is the existence of large, poorly documented, stockpiles of these materials. The US military stockpile of plutonium, for example, stands at 99.5 metric tons, but 111.4 tons have been produced or otherwise acquired by the US over the half century of the nuclear age. The best estimate of the amount removed from the inventory by nuclear tests, radioactive decay, accidents, transfers and losses in production and handling (in the lining of pipes, ventilation ducts, glove boxes, and so on) is 9.1 tons. The remaining 2.8 tons cannot be accounted for (Fetter 1996). No-one seriously believes that it has been stolen; the discrepancy is far more likely to be the result of errors in estimation and book-keeping. Nevertheless, 2.8 tons of plutonium is the equivalent of more than 1,000 nuclear weapons, and uncertainties of this magnitude (uncertainties which are almost certain to be far larger in the case of the former Soviet Union) are clearly very disturbing from the point of view of ensuring that existing stockpiles of fissile materials are adequately accounted for.

However, the 'system dependencies' of nuclear weapons are not restricted to plutonium and highly enriched uranium. Conventional designs of atomic bombs involve an 'initiator' which produces a sudden, intense burst of neutrons to start the chain reaction (see Figure 10.1). The most common design of initiator in unsophisticated nuclear weapons employs polonium. The initiator is crushed by the compression of the weapon's core, mixing the polonium with beryllium: the alpha particles that the former emits in large quantities strip neutrons from the latter. Polonium is made by irradiating bismuth in a nuclear reactor. It was, for example, being made in the pile at

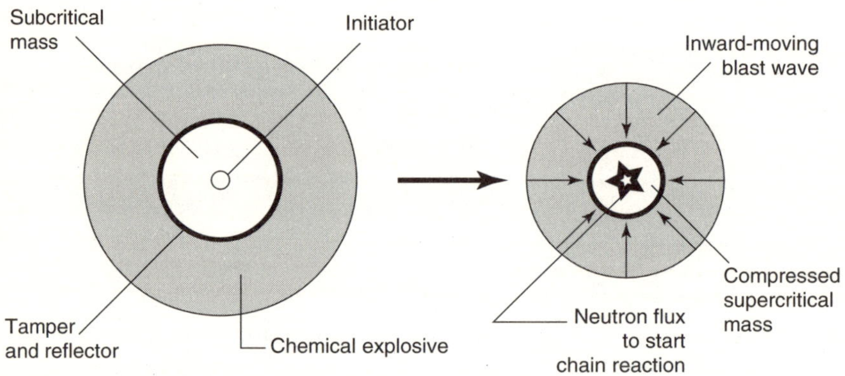

Detonation of a shell of chemical high explosives suddenly compresses a core of plutonium and/or enriched uranium, making the core sufficiently dense to sustain a chain reaction in which neutrons split atoms, releasing large amounts of energy and generating more neutrons which split more atoms, and so on.

In a more sophisticated boosted weapon, the initiator is external, and two gases – tritium and deuterium – are injected into the heart of the core as it implodes. The neutrons generated by their fusion greatly increase the efficiency of the fission chain reaction. An atomic bomb can also be turned into a hydrogen bomb by adding what is called a 'secondary', which contains thermonuclear 'fuel': typically lithium deuteride. Neutrons released by the initial fission explosion bombard the lithium, transforming it into tritium, and the extreme pressure and heat cause the tritium to fuse with the deuterium, releasing enormous amount of energy.

Figure 10.1 A simple atomic or fission bomb of the standard 'implosion' design

Windscale that caught fire in October 1957, in Britain's worst nuclear accident (Urquhart 1983).

Polonium-210 has a half-life[3] of only 138 days, so initiators containing it have to be replaced frequently, and stocks of the substance are of limited value. In part because of this polonium is no longer used in the initiators of sophisticated nuclear weapons. However, these weapons are vulnerable (albeit more slowly) to the decay of tritium, a rare radioactive gas which is used both in their initiators and for 'boosting' (enhancement of a fission chain-reaction by injection of gaseous fusion elements).[4] Tritium's half-life is 12.3 years, so the tritium reservoirs (small pressure vessels containing tritium) of nuclear weapons need replacing after they have been in service for some time. Without replacement, there will come a time when boosting of the hydrogen bomb's fission 'primary' will be inadequate to trigger fusion in the 'secondary' that gives it most of its destructive force. At its most extreme, tritium decay in the primary's initiator might mean no significant nuclear explosion. Tritium decay is already a pressing issue for the US. Since April 1988, it has had no means of producing tritium. The arms reduction process has prevented the situation from becoming a crisis by making available the contents of the tritium reservoirs of decommissioned warheads, but according to Department of Energy estimates that source will become inadequate by 2011.

Tritium, an isotope of hydrogen which exists in nature only in trace quantities, is again the product of large technical systems: a replacement tritium source could cost the US $3,000 million to $6,000 million. Hitherto, tritium has been made by irradiating lithium-6 in specially-modified nuclear reactors, followed by separation in a specialized plant;[5] the cut-off of US tritium production was the result of safety problems with the reactors at Savannah River, South Carolina. The Department of Energy is considering an alternative route, but that is equally large-scale: constructing a large underground particle accelerator, powered by a 350 megawatt electricity supply, the equivalent of that of a small city (US Department of Energy 1995).

It is impossible, on the basis of the open literature, to be sure just how quickly and how completely a sophisticated nuclear arsenal, denied access to tritium, would succumb to the effects of tritium decay. Clearly, the vulnerability of different weapon designs will vary. The only overall estimate I know of is in a 1981 Los Alamos report, which says that a cut-off of access to tritium 'would mean the certain disablement, in approximately a decade, of one half of the [US] nuclear weapons stockpile that uses tritium' (Peabody 1981: 8).[6] However, provided the initiators of nuclear weapons (which seem to require much less tritium than boosting does) keep working, 'disablement' does not mean complete ineffectiveness: at most, it means failure of boosting and failure of the secondary (see Figure 10.1). Unaugmented primary explosions, however, would be enormously less destructive than full thermonuclear explosions. The explosive yield of a modern thermonuclear weapon is typically in excess of the equivalent of 100,000 tons of TNT. An unaugmented primary explosion, however, is likely to be of the order of 500–5,000 tons (Kalinowski and Colschen, 1995: Appendix A4, Collina 1996: 42). These figures correspond to massive explosions (between one-fourtieth and a one-fourth of the Nagasaki bomb), but to a significantly less fearsome threat than full-scale thermonuclear blasts.

Tritium has civil uses that amount to a world market of around 400 grams per year (in terms of quantities, the most important usage is in runway landing lights for remote airfields) (Kalinowski and Colschen, 1995: 140–1). The contents of the tritium reservoir of a single nuclear weapon – around four grams (Mark *et al.* 1988, note 3) – therefore amount to around 1 per cent of the annual civil world market, and so to the kind of quantities whose diversion from civil uses ought to be detectable.

Nuclear weapons as components of technological systems

A nuclear device is not a weapon unless it can be delivered to its target. For terrorist use, transport in a truck or ship might be adequate (the proverbial suitcase bomb would require substantial skills in miniaturization), and for tactical use nuclear weapons have been fashioned into artillery shells, depth-charges and land mines. The main practical delivery means, however, are either aircraft or missiles.

Aircraft are the less demanding in that a heavy bomber can carry and drop nearly all designs of nuclear weapon, at least if one is prepared to take off with the bomb fully assembled and live. However, aircraft can be intercepted, a particularly important drawback if a nation has only a small nuclear arsenal. Outside certain restricted social contexts (most importantly the US Air Force), missiles, especially ballistic missiles, have nearly always been seen as the best means of delivering nuclear weapons to their targets.

Developing a long-range ballistic missile, and integrating a nuclear weapon into it, are demanding tasks of systems engineering. Indeed, the latter, as a distinct engineering discipline, largely began with the first American intercontinental ballistic missile (Atlas) and the first submarine-launched one (Polaris).[7] Non-proliferation policy latched only belatedly on to this 'system' aspect of nuclear weaponry: the current Missile Technology Control Regime was established, at US initiative, only in 1987. Ballistic missiles with relatively short ranges, of the order of 600 km – most commonly the Soviet Scud, or derivatives of it – are relatively widely diffused. Nevertheless, moving from these to an adequate intercontinental ballistic missile, using only indigenous resources, is a demanding task: one, for example, that took China nearly three decades (Lumpe *et al.* 1992).

Missile delivery places important constraints on the designer of a nuclear weapon. There is a strict trade-off between the weight of a missile's payload and its range. The most immediate issue is to reduce the quantity of chemical explosives needed to detonate the weapon, since these are responsible for much of its bulk. This increases the demands for skill in the design and fabrication of an implosion system (which uses simultaneous detonation of high explosives, shaped into a lens structure, to produce an inward-moving blast wave and create a supercritical mass by compression: see Figure 10.1). This, for example, was a major problem faced by Iraqi nuclear weapons designers. Although it seems clear that their government wanted a nuclear warhead for its Scud missiles, Iraqi designers seem to have lacked confidence that they could achieve the requisite compression at the required weight. The design on which they were working when the 1991 Allied bombing interrupted the programme was still too heavy to be carried by a Scud, but the compression of the fissile material in that design was judged by them to be the best they could achieve (and, indeed, they were still some way from practical implementation of that design) (Albright and Kelley 1995: 63).

The need for warhead miniaturization typically grows as one moves from short-range to long-range missiles. The only way to avoid it is to build extremely large missiles (like the first Soviet ICBM, the SS-6), but this has drawbacks. It complicates silo basing and largely rules out what is generally agreed to be the most secure deployment technique: submarine basing. So the miniaturization of nuclear warheads was a key goal in the US, and, eventually, in the former Soviet Union too.

Miniaturization places considerable demands on designers' skills, demands that even in the established nuclear weapons states have not always been met. For example, a warhead with a high yield-to-weight ratio was crucial to the first US submarine-launched ballistic missile programme, Polaris, and had been promised by bomb physicist Edward Teller. Despite imaginative systems engineering (instead of fitting a warhead inside a re-entry vehicle, as had been done in other programmes, the two were designed as an integral unit), designers at the Livermore nuclear weapons laboratory, established by Teller, could not meet the original yield specification. They also discovered in a 1958 test that their design was not one-point safe: a test detonation of the high explosive at a single point (as might occur in an accident) produced a nuclear explosion equivalent in energy to 100 tons of TNT. To make the weapon safer, they inserted a spool of wire into the plutonium 'pit' to prevent an accidental detonation, but found later that the mechanism to withdraw the wire often failed, so many of the first Polaris warheads would have been duds (Spinardi 1994: 55).

More generally, it is miniaturization that makes boosting, and thus tritium, crucial to a sophisticated arsenal. Although more sophisticated mechanical designs and more efficient implosion are important, it is, according to one US weapons designer, boosting that is 'mainly responsible for the remarkable 100-fold increase in the efficiency of fission weapons' since 1945 (Westervelt 1988: 56).

Nuclear weapons and tacit, local knowledge

Technical knowledge is seldom fully explicit. Tacit knowledge – the kind of knowledge we deploy in being able to ride a bicycle without being able to say precisely how it is that we do so – is often important to the successful design and operation of technologies. Sociologists of science, notably Harry Collins, have suggested that even in the most advanced modern science tacit knowledge is largely a local phenomenon. Tacit skills are often not widely diffused, but are the properties of relatively small groups of people, and are transmitted hand-to-hand and face-to-face (see e.g. Ferguson 1977, 1992, Collins 1974).

An emphasis on tacit, local knowledge challenges one of the deepest assumptions about nuclear weapons, held by their opponents as well as by their supporters, that: 'the knowledge required to build nuclear . . . weapons can never be eliminated' (Blechman and Fisher 1994/95: 94). In a stimulating recent article, actor-network theorist Michel Callon suggests the following thought experiment, adapted from Herbert Simon:

> Imagine coloring theoretical statements in red, and all other inscriptions and skills incorporated in human beings and instruments, in green. A Martian contemplating our science from its planet would see a vast green ocean streaked with very occasional and fragile red filaments.
>
> (Callon 1994: 402)

For my current purpose, I want to alter Callon's vivid metaphor, and apply the same red colour both to theoretical statements and to skills incorporated in instruments, while leaving only skills incorporated in human beings as green. What would our Martian now see, particularly if looking not at science in general but at the knowledge needed to design and build a nuclear weapon?

My reasons for adopting the different 'colouring scheme' are portability and mortality. Theoretical statements and instruments are portable and (within limits, as regards the latter) immortal. Skills incorporated in human beings – and tacit knowledge is, quite literally, incorporated, embodied knowledge, as the German expression *Fingerspitzengefühl* ('fingertip feeling'; intuition) reminds us – are portable only along with their human possessors, and share their mortality. If humanly-incorporated skills are needed to design and to build nuclear weapons, and if those skills are local (i.e., concentrated in nuclear weapons programmes, not widely diffused in other activities), then they could be lost as a consequence of a prolonged hiatus in which no new nuclear weapons were designed and built. While nuclear weapons clearly could be reinvented (if it was done once, it can be done again), 'reinvention' might eventually become the correct term: the necessary tacit knowledge would have to be created afresh.

I have explored elsewhere the balance of explicit and tacit knowledge needed to design and build nuclear weapons, and summarize only briefly here.[8] Three types of evidence are relevant: what the original invention of nuclear weapons reveals about the necessary knowledge; what their spread tells us; and what nuclear weapons designers tell us about the knowledge they deploy.

The history of the Manhattan Project suggests that explicit knowledge – at least explicit knowledge in physics – was not, on its own, adequate to permit an atomic bomb to be designed successfully. The physicists involved in the Manhattan Project thought at first that the difficulties lay in producing the requisite quantity of plutonium or enriched uranium, not in turning that fissile material into a bomb. Physicist Edward Teller recalls being advised by future Nobel laureate Eugene Wigner not to join the new laboratory at Los Alamos because its task – designing the atomic bomb – would be too easy. Los Alamos was originally planned to be no bigger than the physics department of a large university. However, it ended up needing several thousand staff, many of whom were engaged in a multitude of apparently humdrum engineering design tasks that the physicists had underestimated. Technological skill turned out to be just as important as knowing nuclear physics.

The spread of nuclear weapons, on the other hand, seems to suggest that the knowledge needed to design an atomic bomb was then highly portable. Six nations, in addition to the US, have successfully exploded nuclear devices, and three others are generally agreed to have – or, in the case of South Africa, to have had – the capacity to do so (see Table 10.1). Other than in the case of Britain (some of whose early designers had taken part in the Manhattan Project) this spread seems not to have taken place by the mechanisms by which tacit knowledge spreads: movement of people or extended face-to-face, hand-to-hand instruction. However, the spread of nuclear weapons is not decisive evidence against the importance of tacit knowledge. The atomic bomb projects subsequent to the Manhattan Project have something of the character of reinvention, rather than copying based upon explicit knowledge. All took longer than the original process of invention (see Table 10.1); all for which details are known were large, labour-intensive projects involving hundreds or thousands of scientific staff; all the cases for which detailed information is available faced a multitude of practical problems; and the possession of explicit information from previous projects did not ease their tasks dramatically.

Certainly, the explicit knowledge generated by the Manhattan Project spread quickly. Soviet intelligence learnt of the West's atomic bomb research as early as September 1941, and a number of agents subsequently gave the Soviets considerable technical data. (Spying, it is worth noting, is an experiment in the nature of knowledge, because spies that remain in place cannot normally transmit tacit knowledge, but only such knowledge as can be

Table 10.1 Approximate chronologies of successful nuclear weapons development programmes

	Start of nuclear weapons development programme	*Date of first atomic test explosion (*) or weapon (†); type of first device*	*Date of first thermonuclear test explosion (*) or weapon (†)*	*Significant personal contact with previously successful weapons design team?*	*Began with attempt to copy previous design?*
US	1942	1945, Pu imp; U gun	1952*	No	No
USSR	1945	1949*, Pu imp	1953*	No	Yes
UK	1947	1952*, Pu imp	1957*	Yes	Yes
France	1955	1960*, Pu imp	1968*	No	?
China	*c.* 1955	1964*, U imp	1967*	No	No
Israel	*c.* 1957 (?)	*c.* 1968 †, Pu imp	? †	?	?
India	*c.* 1964	1974*, Pu imp		?	?
South Africa	1971	1979 †, U gun		?	?
Pakistan	*c.* 1975	1998*, U imp		?	Yes (?)

Notes

a Pu = plutonium; U = uranium; imp = implosion.

b At the time of writing, India and Pakistan have not yet tested, and seem not to have constructed, thermonuclear weapons; South Africa did not develop them; whether Israel has done so is unclear. Pakistan almost certainly possessed atomic weapons well before its 1998 tests.

spoken or written down.) In June 1945, the Los Alamos physicist Klaus Fuchs gave the Soviets a sketch, measurements and a detailed description of the plutonium implosion weapon to be tested at Trinity Site and later to devastate Nagasaki. As they now admit, Soviet bomb designers then set out simply to copy the Trinity bomb. Despite the fact that Soviet physicists were amongst the most sophisticated in Europe, and despite the top priority Stalin gave their work, it took them slightly longer than the Americans had taken to make the original. Producing plutonium in the war-devastated Soviet Union required a massive effort, but also important was the fact that explicit knowledge of the sort passed on by Fuchs and the other agents did not solve all the problems of bomb design. The requisite technological skills had to be created afresh.

Britain, too, began by trying to copy the Trinity/Nagasaki bomb. The task again took longer than the Manhattan Project, led to a design that differed significantly from the original, and involved a multitude of practical problems. For example, the high explosive lenses used to compress the bomb's fissile core had to have a very precise shape. Learning to cast and to machine high explosive with the requisite precision was far from easy, and lenses always seemed to shrink. A humble remedy, PVC tape, was used to fill in spaces and reduce settlement in early British explosive lenses!

More recent nuclear states have had the advantage of sometimes being able to buy key equipment rather than having to make it. But even those whose purchases were most successful, notably Iraq, found that they also needed time to develop the skills to operate the equipment successfully: for example, Iraq was never able to get its centrifuges and electromagnetic separation plant to work satisfactorily.

Understanding of nuclear explosions is most fully explicit at the weapons laboratories of the nuclear powers, which have developed large computer programs, called 'codes', to assist nuclear weapons design. A modern American code can consist of up to a million lines of program, and runs on the world's most powerful supercomputers. Yet weapons designers at Los Alamos and Livermore, interviewed for this research, were adamant that the codes did not eliminate the need for human judgement. The demands of the miniaturization of nuclear weapons, and especially the consequent reliance on boosting, left their job still a difficult art.

None of the above evidence about tacit knowledge is conclusive. For example, even the most detailed historical treatments of the Manhattan Project leave unclear the balance of tacit and explicit knowledge involved in its technological rather than scientific aspects. The duration of projects is a less than decisive indicator of the hardness of the task, because it is clearly affected also by the financial and material resources available. Nuclear weapons designers may well have their own reasons (celebration of their unusual trade; desire to head off a comprehensive nuclear test ban) for emphasizing the continued need in their work for tacit as well as explicit knowledge.

Furthermore, there is some evidence that the tacit knowledge required to develop nuclear weapons is not entirely specific to existing nuclear weapons programmes. Civil nuclear power programmes also demand skills in the handling and machining of fissile materials and in neutronics (the study of the behaviour of neutrons in fissile materials), understanding of which is necessary for designing both nuclear reactors and nuclear weapons. Perhaps even more crucially (because it appears to be the hardest part of the design of implosion weapons, the type preferred by all post-1945 programmes apart from the South African: see Table 10.1), practical experience of sophisticated high explosive systems can be gained in other spheres, notably the designing of armour-piercing shaped charges for anti-tank warfare.

Finally, the demand for tacit knowledge appears to be at its highest in the design of an atomic bomb (or the fission primary of a hydrogen bomb). Given the availability of both a suitable fission bomb, and of the unusual materials necessary to make a secondary (such as lithium-6 deuteride, the standard thermonuclear 'fuel'),[9] then the key to the design task appears to be a piece of explicit knowledge: the Teller-Ulam configuration.[10] The nuclear weapons designers interviewed for this research seemed more confident of the adequacy of explicit understanding of secondaries than of primaries (in particular boosted primaries, where they saw the demand for tacit knowledge at its peak). In addition, the pattern of the spread of the hydrogen bomb is different from that of the atomic bomb, with three of the four states which made the transition from an atomic to a hydrogen bomb doing so more quickly than the US did. Fortunately, there is no known way of making a hydrogen bomb without first having an atomic bomb. So if designing and building the latter demands tacit knowledge as well as theoretical understanding and instruments (and, despite all the above provisos, I feel the balance of evidence suggests that it does), then this remains an important barrier.

The social preconditions of nuclear weapons

Nuclear weapons are not self-evidently necessary or desirable. Only under certain social conditions do they appear so, and the contribution of what I am calling 'structural sociology' to understanding the abolition of nuclear weapons lies in its potential insights into these social conditions. Might one way of ensuring that the abolition of nuclear weapons was irreversible be to eliminate their social preconditions? After all, if these social preconditions were not present, there might be no incentive to seek nuclear rearmament.

For nuclear disarmament to be irreversible, the pre-condition that requires elimination would appear to be the risk of major war between industrialized nations. The 'system' and 'tacit knowledge' perspectives suggest that it may be slower and harder for industrialized nations to recreate nuclear weapons capabilities than is often assumed; nevertheless they clearly *can* do so. In a prolonged, major war between industrialized countries, or in a situation where such a war seemed likely, the chances of an agreement to abolish nuclear weapons breaking down appear substantial.

Is it wholly utopian to imagine a world in which the probability of major war between industrialized nations is low? 'Realist' political scientists point out that these nations' pre-1945 record is not encouraging in that respect, and they attribute the rarity of such war since 1945 to the role of nuclear weapons in making the likely costs of such war outweigh any benefits. Yet it is important to note that this mutual vulnerability of industrialized nations is not the result of nuclear weapons alone. Imagine, for example, the consequences of a bombing raid (using conventional bombs) on a fuelled-up nuclear power station, or reprocessing plant. Such facilities certainly are plausible targets, if only because of their potential role in nuclear re-armament. Furthermore, the changes in war's cost–benefit equation are not simply on the cost side. For example, the mutual benefits of the European Union have been such that most observers would agree that war between those traditional foes, France and Germany, is now unlikely. If the pattern of mutually beneficial interconnectedness and interdependence were to become well-entrenched world-wide, then that would be grounds for hope.

Here we stand on one of the classic divides of social theory. The optimistic point of view, for example advocated in the nineteenth century by Herbert Spencer, could be called 'modernization theory': that as societies modernize and industrialize, trade and democracy spread, and warlike influences diminish. The argument can even be given a feminist twist. If warlike tendencies and male power are associated – and the evidence of this doleful correlation is strong – and if modernization implies a weakening of patriarchy (a plausible, though of course not an entirely self-evident conclusion) then here is another reason, independent of the traditional ones, for optimism.

However, on the other side of the theoretical divide stand not only political-science realism (with its assumption that the behaviour of states is essentially independent of their internal structure) but also Marxism, which, at least in the twentieth century, has emphasized capitalism's role as a cause of war, rather than as a force for peace. There is simply insufficient empirical evidence to be certain whether the optimists or the pessimists are right. Michael Mann, for example, finds no overwhelming evidence that capitalist societies are more militaristic than non-capitalist ones, but equally no clear reason to argue that they are intrinsically more pacific (Mann 1984).

If capitalism's record is ambiguous, there is perhaps better reason to believe the argument, originating ultimately from Immanuel Kant, that democracy is a force for peace. 'Democracies don't attack each other,' declared President Clinton in his 1994 State of the Union address. The matter is not empirically entirely clear-cut: for example, if Wilhelmine Germany counts as a democracy (which is certainly arguable), then the events of August 1914 are clearly an exception to Clinton's generalization; and a near-industry in recent political science has devoted itself to detailed debate about the claim.[11] Furthermore, electorates can sometimes be gripped by war fever, and some of the late twentieth century's minor wars, such as the British re-capture of the Falklands, or the US invasion of Grenada, have been popular electorally. Nevertheless, the enormous costs of full-scale war – even conventional war – between advanced countries are very different from those of recent military interventions in the non-industrialized or less industrialized world. Democracy may not guarantee perpetual peace, but it is plausibly a force for peace.

Actor-network theory

Actor-network theory, however, would regard all the theoretical perspectives discussed in the previous section – political-science realism, modernization theory, 'structural' feminism, Marxism and 'democratic peace' theory – as sharing, for all their diversity, the same theoretical flaw. Each seeks to identify a stable feature of social structure, whether it be the state system, industrialization, patriarchy, capitalism or democracy – and to discover its effects on the propensity to war. Actor-network theory is deeply sceptical about the existence of stable social structure. There are simply no macro-structure constraining events, suggests Bruno Latour, but a constantly open-ended interaction between a multitude of actors. Actor-network theory inherits the interactionist or ethnomethodological insistence that social structure is a precarious outcome, rather than a firm given, and moreover it insists that the outcome is as much 'technical' as it is 'social' in its construction.

The bearing of actor-network theory on armament processes is straightforward, and has for example been explored by Steven Flank in a useful study of the Indian and South African nuclear weapons programmes, and by this author in a study of the development of nuclear missile guidance (Flank 1993/94, MacKenzie 1990). The development of weapons is always heterogeneous engineering, to borrow John Law's phrase: it is always simultaneous engineering of social phenomena as well as of physical ones (Law 1987). A successful weapons programme can indeed plausibly be seen as a network linking physical artefacts and human beings. Weapons system developers have often to spend as much time constructing and maintaining their relationships to human actors (politicians, industrialists, senior officers, the multifarious forms of 'bureaucratic politics') as they do forging physical artefacts.

The usefulness of an actor-network perspective from the point of view of opponents of the nuclear arms race is clear: it suggests a search for the varied and often surprising ways in which the network-building processes of armament can be disrupted (MacKenzie 1990). However, applying the approach to nuclear *dis*armament is a problem of a different order. Actor-network theory prohibits the 'structural sociology' approach of identifying, and then eliminating, the social preconditions of nuclear weapons, for it suggests that the 'social' and the 'technical' are too intertwined, and their interactions too open-ended and too unpredictable, for this to be possible. It would, for example, condemn Marxist pessimism and 'modernization' optimism alike.

This might suggest that the voice of the actor-network theorist in discussions of nuclear disarmament must always be one of extreme caution. If 'everything is uncertain and reversible', as John Law neatly sums up a central aspect of actor-network theory,[12] then there is always the possibility that the ingenious, heterogeneous engineering of a nuclear weapons developer can circumvent whatever technical or social underpinnings of disarmament we seek to create.

On the other hand, actor-network theory's message may not be quite as bleak as that formulation suggests. Unlike realist political science, which believes that states have definite, fixed propensities, actor-network theory asserts that the properties of all actors are not inherent but are relational, so it leaves open the possibility that the very act of nuclear disarmament might change the way states behave (a point I turn to below in discussing the iterative nature of nuclear disarmament). Actor-network theory could also be taken as suggesting the possibility of creating counter-networks (Mort 1995) that might give the abolition of nuclear weapons some stability. Perhaps actor-network theory's greatest virtue is its openness to the way in which 'technical' innovations can transform 'social' possibilities. It was, for example, just such an innovation, the reconnaissance satellite, that helped make possible the partial stabilization of the nuclear arms race in the 1970s and 1980s, a story whose technical dimensions are only beginning to be told (Wheelon 1997).

Of course, it would be naive to expect a 'technical fix' for the problems of making nuclear disarmament irreversible. However, the potential for the development of verification technologies should not be underestimated. The very radioactivity of the key components of nuclear weapons is here their weakness. Plutonium separation, for example, is already readily detectable: the krypton-85 plumes from the reprocessing plants at Sellafield and La Hague can be measured hundreds of kilometres downwind (Weiss *et al.* 1986). Uranium separation is harder to detect, but it is noteworthy that the US reportedly gained evidence of Saddam Hussein's nuclear weapons programme from traces of radioactive isotopes on the clothing of released hostages (Moyland 1997: 18).

Furthermore, verification ought to be seen as a socio-technical rather than as a technical process. Even a modest nuclear weapons programme is likely to involve at least several hundred staff (MacKenzie and Spinardi 1995) and to involve activities that impinge on, and are therefore detectable by, much larger numbers of people. This points to the importance of what Joseph Rotblat calls 'societal verification': systematic measures to encourage and facilitate reporting of violations by citizens of the countries concerned. If 'whistle blowers' are encouraged and protected, and perhaps even rewarded financially (though this last involves the possibility of financially-motivated false alarms), then the achievement of secrecy becomes much more difficult. Actor-network theory's critique of conventional notions of social structure rests on the premise that innovation can make previously large actors small; the converse of this is that small actors (peace movements, non-governmental organizations, even the safeguards branch of the International Atomic Energy Agency, whose annual budget of around $100 million (Moyland 1997: 24) is a tiny fraction of the defence budgets of the major powers) can become large. This leads us back to the question of democracy discussed in the previous section. If governments can punish their citizens arbitrarily, then human whistle-blowing cannot be relied upon (though even North Korea and Iraq have had their defectors), and we must turn primarily to technical means of verification. As democracy and human rights spread, however, reliance upon societal verification becomes more plausible.

Discussion

The bad news

What have we learnt about the feasibility of permanent nuclear disarmament? No individual mechanism for achieving this goal seems likely to be watertight. The route via the control of fissile materials has a reassuring 'physicality', and remains the most important single barrier to further proliferation, but is daunting as a means of ensuring the permanence of the disarmament of the existing nuclear states. Control over new production is relatively straightforward; the major problem is the existence of large, badly documented, stockpiles. Control over polonium and tritium has the advantage that radioactive decay limits the usefulness of covert stockpiles of these materials – relatively quickly in the case of polonium, more slowly for tritium. To the extent that one can be sure that all existing nuclear weapons world-wide depend for their efficacy on tritium, polonium or other rapidly decaying substances, then one can begin to have some assurance that one barrier to disarmament – the risk of it being circumvented by a small number of nuclear weapons being hidden away – may be less alarming than it seems. If disconnected from sources of polonium and tritium, key components of such weapons would eventually waste away.

However, neutron generators not employing polonium exist: the most likely substitute would be a tritium-deuterium generator, which requires much less tritium than boosting does (conceivably as little as ten milligrams).[13] Furthermore, even the much larger quantities of tritium necessary to keep boosting in a small covert nuclear arsenal in working order could be obtained for a long period by secreting a tritium stockpile and purifying it periodically, or conceivably by producing it in a particle accelerator on a scale perhaps compatible with clandestine operation.[14] Furthermore, failure of boosting caused by lack of tritium means only a weaker nuclear explosion, not no nuclear explosion. Even an initiator may not be absolutely necessary. For example, former Los Alamos designer Theodore Taylor has suggested that if a nuclear weapon is built with reactor-grade rather than weapons-grade plutonium (which is difficult, but known to be possible), then the high level of spontaneous neutron emission in reactor-grade plutonium might make it possible to do without an initiator (McPhee 1974).

So there are limits to the extent to which the elimination of nuclear weapons can be made assuredly permanent by controls on the nuclear weapons production system. The difficulty with the requirement for tacit knowledge is that it only slows the reinvention of nuclear weapons: it does not make it impossible. It took roughly three years for the requisite knowledge to be created in the first place in the Manhattan Project. That project had the advantage that, unlike a future covert nuclear weapons programme, it was not in practice constrained by the need for secrecy: there were no German satellites or reconnaissance aircraft able to observe Los Alamos, Hanford or Oak Ridge. Against this, however, must be weighed the fact many of the technologies that had to be created afresh in the Manhattan Project are now available as machines and instruments, and have been much refined. We may perhaps be modestly confident that the need to acquire tacit knowledge afresh through trial-and-error learning means that future programmes (especially covert ones) should not be much faster than the Manhattan Project. There is, however, little reason to expect them to be much slower.

The 'structural sociology' approach to abolishing nuclear weapons by removing their social preconditions is also not watertight. The predictive power of social science is notoriously weak, so only time passing and trust building up can give us confidence that the risk of major war between industrialized countries is sufficiently low. Finally, the strength of actor-network theory – its emphasis on the open-endedness, unpredictability, and socio-technical nature of innovation – is here a source of weakness as well as strength. That small actors may grow, societal verification become possible, and the secrecy required for covert nuclear weapons programmes become unachievable, are all possibilities, not certainties. Actor-network theory is a descriptive, not a predictive, enterprise.

The good news

So none of the four perspectives that I have reviewed – 'system', 'tacit knowledge', 'social precondition' and 'actor-network' – offers a watertight route to permanent nuclear disarmament. The good news, in contrast, is in part simply the diversity of the mechanisms they suggest. To judge the feasibility of nuclear disarmament requires us to judge the efficacy of these mechanisms *as an entirety*. In contrast, the existing literature on the topic tends to consider individual facets of the issue in isolation (often along disciplinary lines), not as a totality.

The other aspect of the good news is that there is evidence that the *interactions* between the 'system', 'tacit knowledge', 'social preconditions' and 'actor-network' aspects of the problem are benign, from the viewpoint of the permanent elimination of nuclear weapons. I have already mentioned one such interaction: between the actor-network emphasis on undermining the socio-technical possibility of secrecy, and the 'social preconditions' emphasis on the spread of democracy. But perhaps particularly important are the interactions between the 'system' and the 'tacit knowledge' constraints on nuclear weapons development.

From the point of view of the demand for tacit knowledge, the simplest kind of atomic bomb is the gun design (in which the critical mass is formed by shooting one subcritical mass into another, using low-powered propellant explosive, rather than the high explosive used in implosion). The gun design avoids the complex detonics and electronics of implosion. The developers of a gun weapon could, furthermore, have reasonable confidence, without a full nuclear test, that it would work. The Hiroshima bomb was used without a full test of its design. South Africa, likewise, felt able to make do with only a laboratory test of its gun weapon.[15]

However, even for nuclear weapons laboratories with years of experience, building a satisfactory gun weapon from plutonium is a demanding task. In July 1944, in the most serious internal crisis faced by the Manhattan Project (its scientific director, Robert Oppenheimer, had to be persuaded not to resign), the plutonium gun was abandoned; and subsequent efforts to build one (pursued as late as the 1970s) have also not come to fruition (Hoddeson *et al.* 1993, Hansen 1988: 21). The problem is that even weapons-grade plutonium has a much higher rate of spontaneous neutron emission than uranium, and in a gun the critical mass is formed only relatively slowly, so the device tends to 'fizzle': to suffer a premature, partial detonation.

Building a gun would therefore in practice mean reliance on highly enriched uranium, rather than plutonium, and so if materials have to be produced afresh, uranium enrichment technology is needed, not just access to a suitable nuclear reactor. For all the progress in enrichment, it remains a demanding technology. Furthermore, greater quantities of fissile material are needed for a gun than for an implosion weapon (in a gun, there is little increase in the density of fissile material, and so critical mass is greater), and gun weapons are harder to miniaturize than implosion weapons. So the problems of delivery are greater for a gun weapon.

More generally, the more sophisticated the delivery system (missile versus bomber; ICBM versus short-range missile) the greater the demands on warhead design. A uranium gun, to be carried by a bomber, can be and has been developed on the basis of explicit knowledge, plus the kind of trial-and-error learning and testing (short of a full nuclear test) that is hard to detect. A relatively simple plutonium implosion weapon to be carried by a bomber also seems feasible without nuclear testing (albeit much harder: the belief that it would be necessary to test an implosion weapon was one reason South Africa opted for a gun design) (De Villiers *et al.* 1993: 102). Moving without a full nuclear test from a weapon to be carried by a bomber to one to be carried by even a short-range missile is harder still (only in 1952, ten years after the start of the Manhattan Project, did the US deploy a nuclear weapon weighing less than the 1,000 kilograms of the Scud's payload) (see the weight data in Hansen 1988). Israel probably has, and Pakistan may have, done this; Iraq would probably have succeeded eventually. In the absence of testing, however, there might be significant doubts that such a weapon would actually work with any reliability.

Making the currently paradigmatic strategic weapon – a miniaturized hydrogen bomb with a boosted-fission primary, to be carried by an ICBM or submarine-launched missile – almost certainly requires retracing many of the steps (including nuclear testing) already taken by the nuclear-weapons states. One significance of tacit knowledge is that even those latter states might have to retrace some of their own steps were they, after a sufficiently long hiatus, to decide to rebuild their arsenals. Blueprints, documents, data and recollections from their original programmes would help, but might not suffice. In particular, as readers of Harry Collins might anticipate, even a detailed blueprint may not allow exact replication of the original weapon:

> Material batches are never quite the same, some materials become unavailable, and equivalent materials are never exactly equivalent; 'improved' parts often have new, unexpected failure modes; different people (not those who did the initial work) are involved in the remanufacturing; vendors go out of business or stop producing some products.
>
> (Miller *et al.* 1987: 3)

Some of these differences would matter, and some would not: knowing whether they would or not is precisely the kind of issue where experienced judgement may be necessary. Miniaturized nuclear weapons (especially those designed to maximize yield/weight or yield/diameter ratios, and to economize on special materials, as much of the recent US stockpile appears to have been[16]) are not a 'forgiving' technology. Designs can be uncomfortably close to what some of those involved call the 'cliff': the region where performance becomes very sensitive to internal and external conditions, with the result, for example, that the explosion of a 'primary' might fail to ignite the 'secondary'.

In addition, not just warheads but the missiles themselves would have to be tested, assuming existing missiles to be destroyed in a disarmament agreement or placed under international surveillance to be used as space boosters. A long-range missile is a sophisticated piece of systems engineering.[17] The many test failures in early US missile work, and the occasional failure of even the most modern space boosters, such as Arianne-5, suggest that if this development has to be started from scratch, without its developers having previous experience,[18] the resultant missiles are unlikely to work with any reliability without reasonably extensive tests – and missile flight-tests, like nuclear explosions, are detectable.

Inter-related system and tacit knowledge considerations also reduce the risk of an agreement to eliminate nuclear weapons being circumvented effectively by secretly retaining some weapons. Nuclear weapons, like all complex technological structures, age – through radioactive decay, corrosion, chemical decomposition and materials creep (Rosengren 1983). These processes are generally slow, but – if a timescale of twenty to thirty years is being considered – important. To my knowledge, no existing design of nuclear weapon could be left unattended that long and then be expected to work with any reliability. Counteracting the effects of ageing requires maintenance of the weapon and replacement of aged parts. Access to the latter (especially to replenished tritium reservoirs) may be a problem; the process may be vulnerable to detection via societal verification; and maintenance and replacement are processes requiring skill. For example, in some of the Soviet weapons inherited by the Ukraine, hydrogen built up dangerously and the Ukrainians found that, despite nationalistic sentiments, they could deal with this problem only by calling in the weapons' original Russian designers. This was a relatively straightforward problem; sophisticated, 'near the cliff' designs may well throw up subtler problems as the decades pass.

Finally, there is an evident trade-off between the size of an arsenal and the visibility of its development. Large-scale production of fissile materials involves isotopic 'signatures' that, as noted above, are vulnerable to remote monitoring techniques, and that might also be detectable via their impact on electric load curves (a 'system' suggestion I owe to Tom Hughes). Even assembly of weapons from covert stocks of fissile materials would risk detection via whistle-blowing and 'societal verification', so again there would be an incentive to keep such work to a small scale. However, the smaller the arsenal, the less likely is it that its possessors would be content with interceptable aircraft or unreliable missiles to carry it, and so the more important would be the system and knowledge constraints on integrating nuclear weapons and missile delivery systems.

Conclusion

The risk that a nuclear weapons abolition agreement would be circumvented by secret development of a large arsenal of sophisticated weapons thus appears to be limited. On the other hand, the risk of the covert development of a small arsenal of relatively crude nuclear weapons, carried by bombers or short-range missiles, is not negligible. Relatively effective defences against such an arsenal are feasible, and it is unlikely to be decisive militarily against a large, widely-dispersed conventional force, equipped and trained to continue to function in the face of limited nuclear attack – a force, in other words, of the sort that the US and its NATO allies currently deploy. However, the potential of even a small, unsophisticated nuclear arsenal as a terror weapon would be considerable. Furthermore, the risks of another form of nuclear re-armament – by either or both parties to a prolonged, major war between industrialized nations – are also not negligible, at least until such time as we can be confident that the probability of such war is very low.

So it seems to me that we have to approach the abolition of nuclear weapons as a *process*: we do not know, and should not pretend to know, now, the solution to all the problems it might throw up (Booth and Wheeler 1992). This is not a reason for delay in starting: actions taken now – for example, to control tritium production world-wide and begin the task of monitoring the tritium stockpile – could greatly ease later stages of the process.

Like almost everything else in our world, from the economic success of nations (Fukuyama 1995) to the stability of empirical knowledge (Shapin 1994), the abolition of nuclear weapons rests ultimately upon trust. As our everyday experience teaches us, trusting is a process: trust is built up through, and sometimes destroyed by, interactions. Trust is also, at least in the context of nuclear weapons, an intrinsically socio-technical matter. If we look only at 'the social', then, unless we are prepared to rely to an undue extent on modernization theory, we can, at best, proceed only very slowly indeed: reasonable certainty that there will be no more war between advanced industrial nations is hardly likely to be gained within the twenty-to-thirty year time scale envisaged here. Similarly, no informed observer can look at the current state of 'technical' verification, and conclude that it is watertight: the avenues for circumventing it are just too evident.

If, however, we consider nuclear disarmament as a socio-technical process, then perhaps – just perhaps – it becomes feasible. The decision to embark on a world-wide process of nuclear disarmament would be, of course, a remarkable declaration of international trust, particularly among the eight overt or de facto nuclear weapons states (the US, Russia, Britain, China, France, India, Pakistan, and Israel). If technical verification efforts, and an increasing scope for societal verification, suggested that all the parties to the agreement were living up to the obligations they had undertaken at a first stage of its implementation, initial trust would then, most likely, be seen as warranted, and the parties involved might then proceed to the next stage of the implementation of disarmament. The very process might help bring about some of its social preconditions, such as the emergence of stable democratic governments in those nuclear weapons states that currently lack them, and an alleviation of the regional tensions (notably between India and Pakistan, and between Israel and its Arab neighbours) that have fostered nuclear armament. There could be no guarantee that an iterative process (Fetter 1996: 40) of this kind would actually converge to the goal of a permanent zero level of nuclear weaponry: a significant breach of trust by any of the parties to such an agreement would almost certainly lead to the others refusing to undertake the next step. Nevertheless, there seems to be a reasonable chance that zero might be obtained, and might be permanent.

Such a process might even find unexpected allies. Compared to the other goals currently being offered to those in the nuclear weapons laboratories, it might prove technically inspiring. Said one nuclear weapons designer: 'Zero in 2005. I can get enthused by that project. What a neat way to cap off your career' (quoted in Weisman 1994: 22). The elimination of all nuclear weapons by 2005 is unrealistic, but major steps in that direction may well be possible by then. Certainly, if the thought of the abolition of nuclear weapons can be pondered in a nuclear weapons laboratory, perhaps the rest of us need to take it seriously too.

Notes

1 The empirical research drawn on here was supported by the UK Economic and Social Research Council, mainly under the Science Policy Support Group Initiative on Science Policy Research on Defence Science and Technology (grant Y307253006), and the interviews were conducted mainly by Graham Spinardi, to whom I am very grateful. This paper is a much-elaborated version of one presented to the conference on Large Technical Systems and Networks, Autun, 27–30 September 1995. I am grateful for helpful comments by audiences there and at subsequent seminars at the University of Keele and MIT.

2 In electromagnetic separation, a gaseous uranium compound travels through a strong magnetic field. The paths of the lighter uranium-235 ions curve more sharply than those of uranium-238 ions, so creating two beams with different isotopic composition.

3 Half-life is the average time interval within which half of any quantity of a substance will undergo radioactive decay.

4 Public attention was first drawn to the significance of tritium for nuclear disarmament by Wilkie 1984.

5 For a description of the process, see Wilkie 1984.

6 All US nuclear weapons now use tritium. Substances other than tritium (such as helium-3, tritium's decay product) might be used in boosting, but 'use of these is considered not to be within reach of present weapons technology' (Mark *et al.* 1988: 116 fn.).

7 See Hughes 1998.

8 See MacKenzie and Spinardi 1995 for the sources of the points drawn on in the following paragraphs.

9 At normal temperatures, lithium deuteride is a solid, so this makes possible a 'dry' hydrogen bomb, in which interactions with neutrons generate tritium from lithium-6. See, e.g., Rhodes 1995: 306.

10 There has been much discussion in the open literature (e.g. Morland 1979, Hansen 1988, Rhodes 1995) of possible details of the Teller-Ulam configuration. Fortunately, there is no need here to add to this discussion.
11 For a useful introduction to this literature, see Owen 1994.
12 Comment made at Keele conference 'Actor-network theory and beyond', July 1997.
13 There are alternatives, such as a plutonium-238/beryllium generator, but the rapid decay of plutonium-238 raises the same problem as with polonium.
14 Fetter (1996: 24) calculates that a 50-gram stockpile of tritium, enough for around twelve weapons, could be maintained by an accelerator with a 200 kilowatt beam power.
15 The test involved making a fissile assembly supercritical, but only for a very short period of time, so that no explosion took place. Richard Feynman compared the original Manhattan Project version of this experiment to 'tickling the dragon's tail' (Hoddeson *et al.* 1993: 347).
16 I do not know whether or not this last point is true of the other stockpiles.
17 This is true of cruise missiles as well as ballistic ones, hence the possible benefits of banning the former as well as the latter.
18 It should be noted that experience relevant to ballistic missile building can be gained from work on space boosters, as the two technologies are very similar.

References

Albright, David and Hibbs, Mark (1991) 'Iraq's nuclear hide-and-seek', *Bulletin of Atomic Scientists* September: 14–23.

Albright, David and Kelley, Robert (1995) 'Has Iraq come clean at last?', *Bulletin of Atomic Scientists* November/December: 53–64.

Albright, David, Berkhout, Frans, and Walker, William (1993) *World Inventory of Plutonium and Highly Enriched Uranium 1992*, Oxford: Oxford University Press.

Blechman, Barry M. and Fisher, Cathleen S. (1994/95) 'Phase out the bomb', *Foreign Policy* 97: 79–95.

Booth, Ken and Nicholas J. Wheeler (1992) 'Beyond nuclearism', in R. Cowen Karp (ed.) *Security without Nuclear Weapons? Different Perspectives on Non-Nuclear Security*, Oxford: Oxford University Press.

Callon, Michel (1980) 'The state and technical innovation: a case study of the electrical vehicle in France', *Research Policy* 9: 358–74.

—— (1986) 'Some elements of a sociology of translation: domestication of the scallops and the fishermen of St Brieuc Bay', in John Law (ed.) *Power, Action, Belief: A New Sociology of Knowledge (Sociological Review Monograph 32)*, London: Routledge.

—— (1994) 'Is science a public good?', *Science, Technology and Human Values* 19: 395–424.

Collina, Tom Zamora (1996) 'Strike up the ban', *Bulletin of the Atomic Scientists* January/February: 41–4.

Collins, H.M. (1974) 'The TEA set: tacit knowledge and scientific networks', *Science Studies* 4: 165–86.

Davis, Jay C. and Kay, David A. (1992) 'Iraq's secret nuclear weapons program', *Physics Today* 45, 7: 21–7.

De Villiers, J.W., Jardine, Roger and Reiss, Michell (1993) 'Why South Africa gave up the bomb', *Foreign Affairs* 72, 5: 98–109.

Ferguson, Eugene S. (1977) 'The mind's eye: nonverbal thought in technology', *Science* 197 (26 August): 827–36.

—— (1992) *Engineering and the Mind's Eye*, Cambridge (MA): MIT Press.

Fetter, Steve (1996) *Verifying Nuclear Disarmament*, Washington (DC): Stimson Center, occasional paper 29.

Flank, Steven (1993/94) 'Exploding the Black Box: the historical sociology of nuclear proliferation', *Security Studies* 3: 259–94.

Fukuyama, Francis (1995) *Trust: The Social Virtues and the Creation of Prosperity*, New York: Free Press.

Gowing, Margaret, assisted by Lorna Arnold (1974) *Independence and Deterrence: Britain and Atomic Energy, 1945–52*, London: Macmillan.

Hansen, Chuck (1988) US *Nuclear Weapons: The Secret History*, Arlington (TX): Aerofax.

Harvard Nuclear Study Group (Albert Carnesdale *et al.*) (1983) *Living with Nuclear Weapons*, New York: Bantam.

Hibbs, Mark (1994) 'Plutonium, politics, and panic', *Bulletin of the Atomic Scientists* November/December: 24–31.

Hoddeson, Lillian, Hendriksen, Paul W., Meade, Roger A. and Westfall, Catherine (1993) *Critical Assembly: A Technical History of Los Alamos during the Oppenheimer Years*, Cambridge (MA): Cambridge University Press.

Hughes, Thomas P. (1983) *Networks of Power: Electrification in Western Society, 1880–1930*, Baltimore (MD): Johns Hopkins University Press.

—— (1998) *Rescuing Prometheus*, New York: Pantheon.

Kalinowski, Martin B. and Colschen, Lars C. (1995) 'International control of tritium to prevent horizontal proliferation and to foster nuclear disarmament', *Science & Global Security* 5: 131–203.

Latour, Bruno (1987) *Science in Action: How to Follow Scientists and Engineers through Society*, Milton Keynes: Open University Press.

Law, John (1987) 'Technology and heterogeneous engineering: the case of the Portuguese expansion', in Wiebe E. Bijker, Thomas P. Hughes and Trevor J. Pinch (eds) *The Social Construction of Technological Systems: New Directions in the Sociology and History of Technology*, Cambridge (MA): MIT Press.

Lumpe, Lora, Gronlund, Lisbeth and Wright, David C. (1992) 'Third World missiles fall short', *Bulletin of Atomic Scientists* March: 31–7.

MacKenzie, Donald (1990) *Inventing Accuracy: A Historical Sociology of Nuclear Missile Guidance*, Cambridge (MA): MIT Press.

MacKenzie, Donald and Spinardi, Graham (1995) 'Tacit knowledge, weapons design, and the uninvention of nuclear weapons', *American Journal of Sociology* 101: 44–99.

Mann, Michael (1984) 'Capitalism and militarism', in Martin Shaw *(ed.) War, State and Society*, London: Macmillan.

Mark, J. Carson, Davies, Thomas D., Hoenig, Milton M. and Leventhal, Paul L. (1988) 'The tritium factor as a forcing function in nuclear arms reduction talks', *Science* 241 (2 September): 1166–8.

MccGwire, Michael (1994) 'Is there a future for nuclear weapons?', *International Affairs* 70: 211–28.

McPhee, John (1974) *The Curve of Binding Energy*, New York: Farrar, Straus and Giroux.

Miller, George H., Brown, Paul S. and Alonso, Carol T. (1987) *Report to Congress on Stockpile Reliability, Weapon Remanufacture, and the Role of Nuclear Testing*, Livermore (CA): Lawrence Livermore National Laboratory, UCRL-53822.

Miller, Marvin and Ruina, Jack (1993) 'The breakout problem', in Joseph Rotblat, Jack Steinerger and Bhalchandra Udgaonkar (eds) *A Nuclear-Weapon-Free World: Desirable? Feasible?*, Boulder (CO): Westview.

Morland, Howard (1979) 'The H-bomb secret', *The Progressive* November: 14–23.

Mort, Maggie (1995) *Building the Trident Network*, PhD thesis, University of Lancaster.

Moyland, Suzanna van (1997) *Verification Matters: The IAEA's Programme '93+2'*, London: Verification Technology Information Centre.

Owen, John M. (1994) 'How Liberalism produces democratic peace', *International Security* 19, 2: 87–125.

Peabody, A.T., Jr (1981) *Some Political Issues related to future Special Nuclear Materials Production*, Los Alamos (New Mexico): Los Alamos National Laboratory.

Pickering, Andrew (ed.) (1992) *Science as Practice and Culture*, Chicago (IL): University of Chicago Press.

Quinlan, Michael (1993) 'The future of nuclear weapons: policy for Western possessors', *International Affairs* 69: 485–96.

Rhodes, Richard (1995) *Dark Sun: The Making of the Hydrogen Bomb*, New York: Simon and Schuster.

Rosengren, Jack W. (1983) *Some Little-Publicized Difficulties with a Nuclear Freeze*, Marina del Rey (CA): R&D Associates, RDA-TR-112116–001.

Rotblat, Joseph (1993) 'Societal verification', in Joseph Rotblat, Jack Steinerger and Bhalchandra Udgaonkar (eds) *A Nuclear-Weapon-Free World: Desirable? Feasible?*, Boulder (CO): Westview.

Rotblat, Joseph, Steinerger, Jack and Udgaonkar, Bhalchandra (1993) *A Nuclear-Weapon-Free World: Desirable? Feasible?*, Boulder (CO): Westview.

Shapin, Steven (1994) *A Social History of Truth: Civility and Science in Seventeenth-Century England*, Chicago (IL): University of Chicago Press.

Spinardi, Graham (1994) *From Polaris to Trident: The Development of US Fleet Ballistic Missile Technology*, Cambridge: Cambridge University Press.

Taylor, Theodore (1993) 'Technological problems of verification', in Joseph Rotblat, Jack Steinerger and Bhalchandra Udgaonkar (eds) *A Nuclear-Weapon-Free World: Desirable? Feasible?*, Boulder (CO): Westview.

Urquhart, John (1983) 'Polonium: Windscale's most lethal legacy', *New Scientist* 97 (31 March): 873–5.

US Department of Energy, Office of Reconfiguration (1995) *Final Programmatic Environmental Impact Statement for Tritium Supply and Recycling*, Washington (DC): Department of Energy (DOE/EIS-0161).

Vartabeian, Ralph (1995) 'US launches race to save nuclear arms know-how', *Los Angeles Times*, 28 August.

Weisman, Jonathan (1994) 'Early retirement for weaponeers?', *Bulletin of Atomic Scientists* July/August: 16–22.

Weiss, W., Stockburger, H., Sartorius, H., Rozonski, K., Heras, C. and Östlund, H.G. (1986) 'Mesoscale transport of ^{85}Kr originating from European sources', *Nuclear Instruments and Methods of Physics Research* B17: 571–4.

Westervelt, Donald R. (1979) 'Can cold logic replace cold feet?', *Bulletin of Atomic Scientists* February: 60–2.

—— (1988) 'The role of laboratory tests', in Jozef Goldblat and David Cox (eds) *Nuclear Weapon Tests: Prohibition or Limitation*, Oxford: Oxford University Press.

Wheelon, Albert D. (1997) 'Corona: the first reconnaissance satellites', *Physics Today* February: 24–30.

Wilkie, Tom (1984) 'Old age can kill the bomb', *New Scientist* 16 February: 27–32.

11

SOME THEORETICAL AND CRITICAL REMARKS ON THE AIR TRANSPORTATION SYSTEM AND OTHER LARGE TECHNICAL SYSTEMS

Alain Gras

One may wonder whether large technical systems (LTSs) belong to a linear evolution of technological development, thereby representing the last stage (to date) in a progress-oriented process, or whether they are a radically distinct phenomenon, politically, ideologically and technically. In my view the second alternative is the right one: LTSs constitute a unique object for the sociology of technology, because they are an essential feature of 'modernity' and because they support its world-wide expansion. Above all, they reinforce the obsession with grandeur and power which currently guides the prevailing technological trajectory.

I will first attempt to throw light on this 'phenomenon' and describe its specificities by discussing two ambitious studies of 'big technology'. I would like to demonstrate that a nuclear reactor, for example, is *not* a large technical system, although it certainly is a big technology that can be regarded as a system, in the common sense of the term – as can any technological artefact –, and although it is a part of electricity supply LTSs. Even more important: although all LTSs include a network as one of their constitutive elements, not all networks are part of LTSs: road networks, for instance, do not (yet?) belong to any LTS (see however the discussions in Grundmann 1994, Juhlin 1994, Parrochia 1993, Offner 1994).

In the first part of this chapter, therefore, I shall discuss some theoretical questions about LTSs. I shall do so by criticizing the implicit philosophy of history of two authors who have taken an interest in the history of big technology, Maurice Daumas and James Beniger. In the second part I shall examine the system of civil aviation, which will allow me to 'revisit' some key LTS notions: the notions of 'reverse salient', seamless web, overload, control.

I shall argue that the development of this technologically innovative LTS is blocked by a socio-technological reverse salient combining congestion problems and the limitations of current methods in air traffic management.

Networks: information, control and the rise of the large system

Networks explain nothing: Caesar's horse and the control revolution

The author of an impressive history of science and technology, Maurice Daumas criticizes Raymond Aron's claim that it took Caesar approximately the same amount of time to go from Rome to Paris as Napoleon (Daumas 1991). Aron supports a theory of the irregularity of technological progress that Daumas refutes. Daumas's argument that *Lutetia Parisorum* ('Lutetia of the Parisians') would have interested Caesar less than Reims (which was located on the road to Britain) can be easily rejected: obviously, Aron could have written 'from Rome to Reims' without altering the meaning of the metaphor. But Daumas's commentary turns against him: flows through space are not determined by physical geography but by the processes through which flows are transformed into physical reality, in this case roads and horsemen. The shape of a network is always connected to political, economic, military and/or religious concerns: they all determine the morphology of flows by giving them a reason to exist.[1]

Historians of technology often display a surprising absence of critical judgement in my view. For example, they rarely question why certain civilizations did not use the wheel despite the fact that they were aware of its existence. The possibility that the wheel – a very useful object for communication – might not have been necessary in their system of values, or that its diffusion would have occasioned dangerous changes in the prevailing 'social balance', does not occur to them. Likewise, the deterioration of roads during the Merovingian period could be interpreted not as some sign of loss or decline, but as the emergence of another 'social fabric', the birth of a new social system and not a step backwards.

In order to avoid the determinism underlying most conventional history of technology, the sole solution lies in the principle set forth by Michel Serres, Trevor Pinch and Wieber Bijker: it is not the failure of artefacts but their success which should be explained. I shall apply this principle below, but I would first like to give examples of the main danger by which social studies of technology are threatened.

According to Daumas, 'since the beginning of humanity technical progress has evolved in a regular manner with relatively few ruptures' (Daumas 1991: 4). I criticized this view long ago (Gras 1979, 1980). It is an illusion – but a widely shared one – which excludes all records of past acts and facts which do not fit into modern categories and value systems as well as many civilizations (most of them in fact) which share the central belief, exactly opposed to ours, that technological progress is neither necessary nor inevitable.[2]

In fact, efficiency is not an objective notion against which all 'technical facts' can be assessed, and Aron's statement is not concerned with technology, but with the social system. Aron simply claims that the 'imaginaire' of speed and transportation modes remained relatively static until the eighteenth century: even if Napoleon's civil engineers were better equipped than Caesar's, the obsession with speed had not yet taken hold at the time, nor had it yet transformed the landscape to satisfy the modern desire always to be somewhere else. (See Berthoud and Busino 1991, Studenyi 1995 and Offner's contribution in this volume.)

The net needs information and control to work

James Beniger (1986) also discusses the genesis of networks and big technology. He argues that the invention of computers is a natural continuation of the 'control revolution' already underway in the nineteenth century. Beniger, as a matter of fact supports the idea that the control revolution happened after three others: the neolithic, commercial and industrial revolutions. He thus suggests a sort of natural history: 'information processing develops subsequent to the extraction and processing of matter and energy . . . processing and distribution require control, under this model, and control depends on information services' (Beniger 1986: 33).[3] This very pertinent comment shows how essential are both the circulation of flows and the capacity to be informed about these flows.

Beniger then proceeds to make a detailed analysis of the manner in which the new LTSs brutally modify the relations between man and machine. He clearly describes the difference with the old way of relaying within networks:

> The conductors [of the Western Union Railroad in the 1850s] take on new significance: they are possibly the first persons in history to be used as programmable, distributed decision-makers in control of fast moving flows through a system whose scale and speeds precluded control by more centralized structures.
>
> (Beniger 1986: 225)

One must, however, put this scenario back into a broader history of changing mentalities. Information is a very recent concept, as Beniger himself acknowledges. The preceding quote shows one of the reasons why 'checklists' appeared.[4] But this is a temptation to look for a 'cause for a cause'! And when Beniger looks for the origin of the control process in the amoeba, he wipes out in one sweep the discontinuity which he had introduced with his theory of revolutions. This point of view reintroduces a human-oriented direction of time and threatens to negate the interesting thesis that the historical invention of the grouping (system + network + control) is constituting modernity, in a non-linear process of evolution. Philosophical historicism comes back in through the window!

How can the pitfalls of 'oriented time' and technological determinism be avoided? Let us consider in greater detail the case of the Roman horseman. Whenever he went for a ride on Via Domitia he got information about where he was going, and could check his plans before he went. However, he made his decision based on local time and while he obeyed orders, these orders did not indicate a procedure to follow. His personal history was independent from the horsemen's collective organization. The management of horsemen did not consist in the close surveillance of flows, but in the loose control exerted over the environment, which in that case consisted of *emporia*, i.e. stores where horsemen could pick up supplies along the way.

Getting and checking information are socio-technical facts that have always existed, but it is their insertion in a new form of flow management where checking is centralized – or rather 'delocalized' from the point of view of the flow unit (horserider, pilot, driver, electron, manager, etc.) – that constitutes the radical novelty of large technical systems. One could imagine Caesar's horseman fitting into an LTS if this horseman carried a mobile phone that connected him to the staff headquarters in Rome. However, for this to happen we would have to endow Roman senators with the mentality of American businessmen, which is a negation of history!

Daumas and Beniger made a mistake which is common in the history of technology: they confuse the function of the (useful) object with the object itself as a tool and know-how (see Mauss 1936, unfortunately not translated into English). The same material object may have very different meanings and functions depending on symbolic contexts. And, the control exerted by a biological cell or by Caesar's soldiers are about as connected to the control of modern LTSs as a gas-powered chainsaw is connected to the paleolithic stone axe!

This naturalist sociology – or this naïve historicism in the case of Daumas – are just plain wrong with regard to LTSs. They must be left out of our theoretical framework, and the only plausible hypothesis is none other than that contained in Michel Foucault's *Les mots et les choses*: if the 'arrangements' (as understood by that author) allowing for the emergence of this new type of 'macro-systemic' organization 'eventually disappeared as they had appeared . . . we can bet that [this way of being-in-the-world of] humanity would be erased like the tide takes a sandcastle off the beach' (Foucault 1974: 398).

In sum, LTSs can be considered as a new radical element born of industrial society. Nothing more. However, one must then trace the genealogy of this new technological form – genealogy in the sense that philosophy has given to this term since Nietzsche, i.e. the description of the origins as a device (or as an 'arrangement' in the words of Foucault) that allows a given fact to exist both in concrete and imaginary terms. The origins are therefore not thought of as a historical train of causes and effects but more like the violent agglomeration of elements which each follow their own history and which crystallize in a new ideological grouping (Caron 1985, Kaijser 1995). As a consequence, we assert the emergence of a radically new social phenomenon, the novelty of which does not lie in the network, but in the information system and the rational, centralized organization of flows. And to emphasize this radical novelty I shall hereafter use the phrase 'large-scale technical system' (LSTS) instead of LTS, as an equivalent of the French 'macro-système technique' (Gras 1993, 1997).

The genealogy of LTSs: a drive for energy and power?

Technoscience no longer limits its aim to 'becoming the possessor and master of nature' (Descartes), but exerts its power on every aspect of human life. LSTSs belong to this relatively new vision of the world. One might wonder why the obsession with the use of power – free of all ethical constraints and applicable in all places and at all times – became a value (Balandier 1985). We know that non-occidental or pre-modern civilizations had the opposite obsession, that of the risks involved in leaving power to the discretion of men. There is no answer to this question. Sociological truth is like a strange loop, the loop of existence itself, where one encounters, as Heidegger would say, the 'questioning of the question'. The only definite fact consists of the appearance of a new relation to the world which only Mumford's mega-machine could have anticipated – albeit without his perceiving the new possibilities opened up by the harnessing of energy (MacNeil 1982, Salomon 1992, Janicaud 1995, Latouche 1995).[5] This new paradigm rests on a tripod where mobility and energy combine (see Figure 11.1).

If we consider a locomotive that uses coal for locomotion while transporting the raw material from the site of a mine, we get a clear archetypal image of this new geography of energy flows. A simple example of this new 'imaginaire' can be found in the common economic discourse: indicators used by economists to characterize levels of development – such as the 'gross national product' – are indicators of flows! Without this assumption, I find it difficult to understand the insertion of the macro-system into its ecological niche.

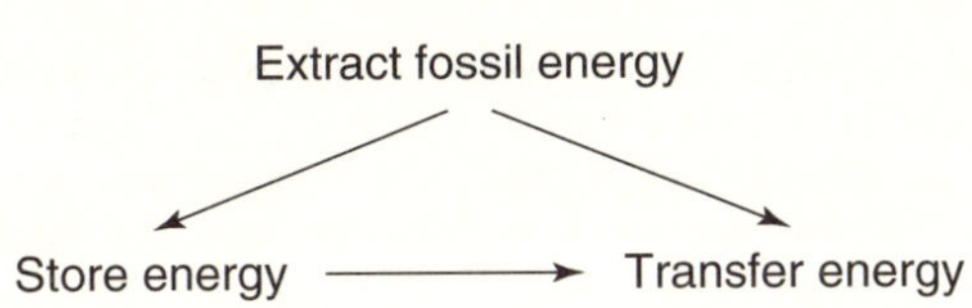

Figure 11.1 The energy–mobility tripod

Weaponry is typical of this play with time and space to gain power – for the purpose of destroying the power of the enemy. Table 11.1 shows that the rise of destructive power combines with the rise of weapon mobility: from the arrow to the missile. This rise is recent, and did not take on such an accelerated form until the middle of the nineteenth century, at precisely the

Table 11.1 The evolution of armament

Weapon	*Lethality index*
Light weapons	
Sword	20
Javelin	18
Bow and arrow	20
Crossbow	32
Arquebus, 16th century	10
Musket, 17th century	19
Flintlock, 18th century	47
Rifle, mini bullet, middle of the 19th century	150
Rifle, magazine, First World War	780
Machine gun, Second World War	18,000
Modern light machine gun	21,000
Automatic grenade launcher, High explosive fragmentation grenades	1,500,000
Heavy weapons	
Cannon, 16th century	43
Cannon, 17th century	230
Cannon, 18th century (Gribeauval)	4,000
75 mm cannon, shell	34,000
Howitzer, Second World War	660,000
Tactical missile, nuclear head of 0.05 kg/t	60,000,000
Modern weapons (examples)	
Main battle tank	3,200,000
Heavy bomber, with high explosive fragmentation grenades clusters	210,000,000
Strategic guided missile, with nuclear head of 25 megatons	210,000,000,000 (highest index to date!)

Source: Perry Robinson 1978.

Note Lethality index measures a combination of variables: destructive power, mobility, fidelity, number of targets per shot, etc.

According to the author of the article, the comparison between the capacities of the new conventional weapons and the capacities of smaller calibre nuclear weapons shows that the gulf between them has been partially filled in. This makes the rules of the game and the balance of terror obsolete. After the recent nuclear experiments of China, India and Pakistan this question is still with us.

moment when the West, in another power move, imposed its law on the world (Braun and Joerges 1994).

In comparison, LSTSs seem to be a peaceful expression of this evolution. However, the playing out of power struggles makes this peacefulness ambiguous. Arden Bucholz (1994) even evokes a 'Prussian Paradigm' in his description of the extraordinary growth of the fire power of the Prussian armies between the war against Denmark in 1864 and the war against France in 1870, due to the rational use (in the military value system) of the railway LSTS. Within six years, there was a fourfold increase in the capacity to mobilize and send recruits to the front! Later, from the beginning of the First World War, 560 convoys of fifty-four carriages crossed the Hohenzollernbrücke every twenty-four hours! Likewise, the development of aviation technology was boosted by its use in the two world wars. I would argue that, had the First World War not occurred, civil aviation would exist today only in its embryonic form, and the air transport LSTS would have been based upon the technology of the dirigible.

A successful LSTS: civil aviation

The setting up of a world-wide civil aviation network took place at the end of the Second World War, with the creation of the International Organization of Civil Aviation (IACO), the International Air Transport Association (IATA), and later of the IFALPA for pilots, the European Convention for Civil Aviation (ECCA), etc. These organizations supported a process of globalization. For instance, the English language became the single official language in use in civil aviation (although Russian and French are also official languages within IACO). Member states accept international regulations and follow IACO prescriptions. Companies design competition rules as well as fair conduct rules within IATA: LSTSs also imply a philosophy of life!

Of course, this international network developed unevenly. In France, for example, the modernization of infrastructures and institutions occurred only after 1948, when a BOAC Dakota crashed because it lacked the appropriate guidance systems for landing. It was at that time that the transport minister, Jules Moch, decided to create an engineer corps, in particular a corps of air navigation engineers (chosen from among the students of the prestigious Ecole Polytechnique), a corps of air traffic controllers and a school to train them, the National School of Civil Aviation (ENAC). We see here a common characteristic of LSTSs: accidents are used to level up the parts of the system that lag behind. From the beginning, the air transport LSTS converted disastrous events into useful data. The first department of accidentology, the English AIB was created as early as 1915.

In fact, LSTSs always have a 'bureaucracy of memory', which uses information technologies to replace accidental events in a long-term perspective (Weingart 1989). Charles Perrow's (1984) interpretation of accidents as being 'normal' in these kinds of organizations should perhaps be turned upside down: accidents are normal in a LSTS, not because they signal the – unavoidable – failure of the organization that sustains the system, but on the contrary because they allow it to keep pace with the evolution of its environment. Obviously, for this to happen, the system must have perfected a way of treating accidents that allows knowledge to be gained along with limiting damage.

After this historical perspective, I would like to discuss some key notions of LSTS theory: the notions of 'reverse salient', 'seamless web', overload, and control. This discussion will allow me to conclude by discussing the sociological nature of LSTSs.

A reverse salient: the sound barrier as a social problem

The image of the reverse salient was borrowed by Thomas Hughes (1983) from the military vocabulary of the First World War. In France one obviously thinks of Verdun, which was a reverse salient for both parties in turn. The notion is without doubt systemic, because it brings out the interdependences between the parts of a technological ensemble. There have been many instances of reverse salients in the air transport LSTS since the Second World War. A famous one is the sound barrier that greatly affected the commercial development of the Concorde supersonic plane. The Concorde would appear to have surmounted the sound barrier, but in the end it will go down in history as a prototype made in several copies.

In 1969, when Henri Ziegler – the founder of Airbus Industry, the major European aircraft building company – predicted that 'hundreds of Concordes' would be required to meet demand in 1980, he shared the common illusion that the quality of the product would generate demand, and that 'technological push' would impose its law. It had not occurred to him that this kind of evolution would encounter another trend: a new awareness of nature. Ecological movements then 'invented' a reverse salient: noise. The US Federal Aviation Administration (FAA), supported by the US aircraft industry (and soon followed by IACO), rapidly banned the 'supersonic bang' from the continent: Concorde would not serve on the busy line between New York and Los Angeles, and it also had considerable trouble being admitted into New York's airport. Other problems like Concorde's small size and the voracity of its engines, contributed to the termination of its commercial career.

Thus, the air transport LSTS has not conquered the sound barrier to this day. This is not just an engineering issue: the sound barrier is a political, industrial and certainly a commercial issue. In reality, reverse salients are part of what Marcel Mauss called a 'total social phenomenon'. Even though the expression 'reverse salient' suggests a technological perspective, the phenomenon has different meanings depending on the level of interpretation. And what has become of the sound barrier reverse salient today? It has disappeared from the scene, but it still rules the game: the supersonic post-Concorde is tucked away on engineers' shelves, and no one dares envisage its production, even in the long run.

In the meantime, engineers have chosen another path for technological innovation: the introduction of micro-computers in the cockpit – now called a 'glass cockpit', because of the replacement of traditional instruments by computer screens. This path of innovation brings out a new technological trajectory on which other reverse salients could appear – but not the sound barrier (see Gras *et al.* 1994). To overcome this reverse salient, the front line was displaced, and another battlefield chosen, while everyone kept quiet about the defeat. It is not a techno-logical choice, but clearly a socio-logical one.

The seamless web and the virtual airspace

The air transport LSTS gives a near perfect illustration of the 'seamless web' as a mixture of technological and social facts. In general, air traffic is faced with two constraints: first, it is impossible to interrupt the movement of flying objects; second, (historically produced practices imply that) air traffic efficiency and safety are obtained through the continuous, centralized control of flying objects and by the compliance to strict distance rules (Maignan 1991). Control over flying objects is ground based, and this requires that the airspace be divided into 'regions' in which flows are monitored and steered by air traffic controllers. This brings in a cognitive constraint: considering that the human brain is limited in its capacity to assimilate information, one single individual cannot follow more than fifteen or twenty planes at any time, or deal with more than forty planes per hour.

In addition, because of their high speed, planes must be kept at a large distance from each other, and therefore the 'virtual size' of an aircraft in the air is considerably bigger than its ground size: it can represent a volume about 10 km in length, and 600 m in height and width. Hence, air space is quickly saturated, even if to the naked eye it looks empty. Congestion results both from the cognitive limits of the air traffic controller and from the distance rules, which are two completely human factors.

The sectors system is intrinsically complex. France, for example, is divided into five regions of control which are in turn divided into 'sectors'.[6] It is interesting to note that the shape of these sectors is never a simple geometric one. Note also that distance rules (and hence the 'virtual size' of planes) vary widely along the way: this is because they depend on the characteristics of ground equipment in each sector (see Figure 11.2).

In Europe it is not very efficient to reduce the speed of planes in flight (the American model; see La Porte 1988), because distances are short. Rather, planes are simply denied clearance for take-off when congestion is anticipated at destination or somewhere on the route. For example, when southern Europe

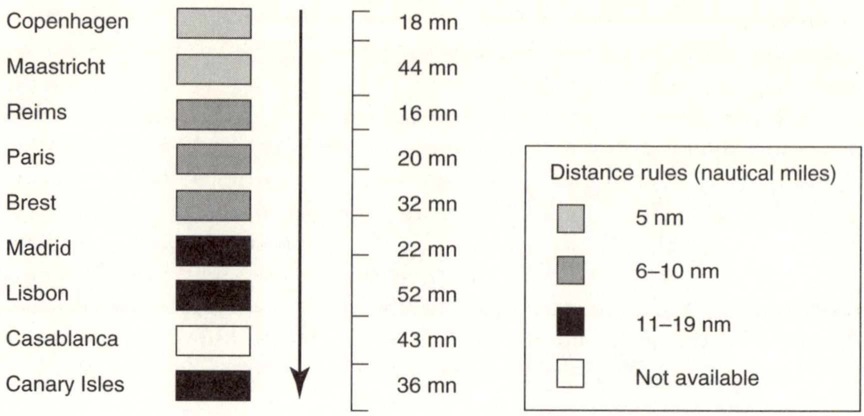

Figure 11.2 Distance rules and flight durations in air space sectors: the example of a flight from Copenhagen to the Canary Isles

Note Right hand scale shows time in minutes (mn) aircraft occupy the relevant air space.

receives charters from Great Britain and northern Europe, airports at destination may not always be congested, but in the Brest sector criss-crossing is intense. For this reason, the air traffic controller in Brest must limit departures from Copenhagen, Frankfurt, or Paris (cf. Figure 11.3).

A certain number of planes are authorized to take off every hour and each one is allowed a margin (fifteen minutes) for departure. If they exceed it they run the risk of having to wait at least one hour for another slot. But flight times not being the same, lag times must be taken into account: what is a quiet hour in Copenhagen, for example, 6:00 a.m., becomes a busy hour when the plane flies over Paris at 7:30 a.m. or when it enters Madrid's sector an hour later. The resulting situation is often paradoxical, with an empty sky or an empty runway, but no clearance for take-off. Pilots who do not accurately conceive of the complexity of the system, bitterly complain about it; and passengers are unable to explain apparently absurd delays which they often attribute to apathy, while it is in fact the authoritative management of flights that allows the air traffic 'load' to be evenly distributed along the arcs of the network.

Specific difficulties also result from the fact that flows are not transferable to other lines instantaneously, as can be done on an electrical energy network. In Europe, where distances are relatively short, the only solution is to control flows at their source, which means keeping the plane grounded if necessary. But this requires complex organization between the airports to find out which airport, which flight and which hours are the most concerned. The control system has recently been further centralized. It is now the task of the Central Flow Management Unit (CFMU) in Brussels to establish schedules for the whole of European air space. The mastery of future time is inscribed in a

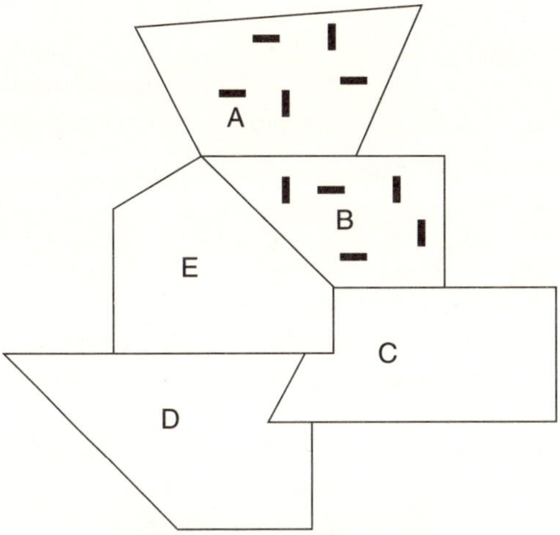

Figure 11.3 'Remote congestion'

socio-technical tool, the regulatory cell of the CFMU that plans flights from six months (strategic planning) to an hour (planning tactic) in advance.

With the steady increase of air traffic (expected to exceed 7 per cent in 1998), it is not certain that this organization will be successful. The future depends on the balance between 'tight' and 'loose' coupling or on the level of determinism that can be maintained. Behind the apparent authoritarianism of centralization there may well be many hidden ways to change the rules, especially at the level of 'correspondents' playing the role of intermediary or communications agents in the flight management centres. The delocalization (centralization) of decision making and the extension of the 'seamless web' reach their limits, thus questioning the principles of air traffic control.

Overload and institutional decision rules

Security rules for flying and the organization of air traffic control are important constraints faced by the air transport LSTS. But to match the rapid growth of demand, the system has to cope with many other constraints. For example, increasing the number of runways and airports will not only fail to solve the problem of air traffic sectors, but it would necessitate interfacing with many other systems such as, rail, road and energy supply – and cope with the opposition of local residents.

More runways mean more terminals, as well as a proportionate increase in the number of personnel on the ground for reception, security, police and customs. Imagine the traffic jam caused by four jumbo jets taking off within a few minutes of each other: this means 2,000 people going through security, with at least 2,000 pieces of carry-on luggage that have to go through x-ray, numerous body searches, etc. (recall that, unlike trains, airplanes cannot leave before every last registered passenger is in the cabin).

If more airports were built, they would have to be located farther and farther away from major urban centres, and surface transportation would be needed to take travellers to their destinations. New infrastructures would have to be created – which would undoubtedly clash with the expectations of the local residents and of environmentalists. Moreover, passengers would have to change transportation modes several times in order to reach their final destination.

From this brief account, it is easy to understand that the air transport LSTS is confronted with an internal contradiction: passengers expect planes to take them to their final destination – which planes are less and less able to do. Of course, it could be argued that the problem of overload may be interpreted as a reverse salient. This would mean assuming that congestion calls for a purely technical solution. One such solution was pursued: automating control, or at least a part of the air–ground connection. But despite fantastic promises from artificial intelligence programs, like that of the Massachusetts Institute of Technology, and the gigantic sums of money sunk into the AAS program (Advanced Automated System) in the US, it was a complete failure. It does not work much better in Europe.

Charles Perrow (1984) noted that, since 1960, air traffic control constitutes 'a striking example of the possibility of reducing complexity and coupling in non-transformation systems', but he adds that the US Federal Aviation Administration (FAA) 'is pressing for more automation, thereby reducing the numbers of controllers extensively. Both of these, I would suggest, will lead to much tighter coupling – that is less recovery from incidents' (Perrow 1984: 159). Perrow was right because we know now, more than a decade later, that the AAS project of the FAA was misleading. The cause for this failure can no doubt be found in the will to force reality into a deterministic mould for this type of automatization: the FAA had not understood at the time that civil aviation is a social system and that air space is a virtual space which would fall apart if not held up by human beings standing on the ground.

However, to bypass the complexity of the ground system, serious consideration is being given to the idea of *free flight* – meaning that, to a certain extent, planes would no longer have to follow predetermined air routes and to pass through fixed points (Wise *et al.* 1992). The pilot would take the plane straight from one airport to another by the path he considers the most appropriate, and decide for himself on which manoeuvres to perform to avoid other planes (especially at high altitude). 'T-CAS' (on-board anti-collision radars) allow the surrounding sky to be viewed: with adequate procedures, the pilot could then act as an automobile driver individually responsible for avoiding other vehicles.

If free flight were implemented, it would be an example of where the LSTS regulates itself and increases its capacity by destroying a part of its own control organs. The problem is no doubt more complex than that because regulation of air traffic will always be necessary, but it could lose its strategic importance. The important point here is that actors in the air transport LSTS have different views about the possible evolution of the system. These views, as well as any technological innovation in air traffic control, can be related to one of two conflicting ideals: the ideal of free flight (the 'Icarus' ideal) and the ideal of fully controlled flight (the 'mechanical bird' ideal) (Gras *et al.* 1994). Table 11.2 summarizes the two ideals.

Table 11.2 Icarus and 'mechanical bird' ideals

'Icarus' ideal	*'Mechanical bird' ideal*
The latent structure of technological projection between air and ground:	
Decision-making left to the pilot Perception of ambient space and a large part of the environment by the crew Contingency rule for collision avoidance in the air (and possibly air corridors) Simple approach guidance, etc.	Plane guided and regulated throughout the flight Compulsory routes with flight levels decided on the ground General flow control Fully controlled aircraft, etc.
This involves technological endeavour focused on:	
Synthetic visualization of the environment and simplified representation of reality (airborne radar, anti-collision instruments such as T/CAS, etc.) Easier communciation between aircraft A tightly regulated code of procedure governing the relationship between the planes and the pilots who fly them A statistical concept of control rather similar to that used for road traffic, etc.	Collection of information in flight as well as before the flight Standardization, homogenization and expansion of equipment capacity Direct air–ground dialogue via the machine and development of airborne equipment in direct liaison with the ground, etc.

Conclusion: technological blockade and sociological solution (in an evolving LSTS)

The study of civil aviation clearly demonstrates that neither reverse salients nor the 'seamless web' of air traffic control nor congestion are purely material facts. All are socio-historical facts with technological content. It is striking that the further development of civil aviation should be now held up by the very processes and choices that made possible in the past the massive increase of traffic (under conditions of remarkable safety). Furthermore, potential technological 'progress' will not solve the problems linked to the systems' interdependencies (Rochlin 1996): airplanes will not fly faster (only its range of action will increase); the 'glass cockpit' will not modify the system's equilibrium (it leaves the pilot 'in the loop', and automation cannot be envisaged); safety standards will be maintained as much as possible, but not improved; the automation of air traffic control will not be realized – at least not in the next twenty years.

Yet air traffic will continue to grow, and solutions will be provided to future reverse salients, but they will not be technological ones. Technological innovation will only offer new instruments to accompany 'ideas' that will develop in the Icarus–Mechanical bird space. The example of the air transport LSTS should convince the reader that, at any time in history, 'the way is open'; that radical changes which will seem natural *a posteriori* in fact result from subjective choices; that technology is therefore not autonomous (self-contained); and ultimately, that even when faced with dinosaurs such as large-scale technical systems, human beings remain free.

Notes

1 The pilgrimage to Compostella, the most famous of the Middle Ages, had a funnel-shape road network, with its narrow end located at Saint Jean Pied de Port on the Basque Spanish border.

2 For example, Daumas sees the invention of the stirrup as a 'progress' but does not explain why native Americans rejected it. See also Feenberg (1995) about the 'Japanese game' with technology.

3 I was surprised to read (p. 32): 'an earlier materialist historian, Karl Marx, made the same point'! Harvard humour?

4 A look at the question of procedures shows how techniques and society are inextricably linked. A checklist assumes an absolute adherence to the rules. It belongs to a period of bureaucratic rationalization which accompanied the advent of industrial society, as well as to a culture of 'law and order' (like that of the Western Union). The puritan ethic, the inner directed personality according to D. Riesman, obviously underlies this way of working. In 'Latin' or catholic countries people believe that the rule is the expression of a power whose legitimacy is always in question. People conform because they have to. This question is very important for the use of new technologies in various civilizations (Scardigli

1993). Rules in technical procedures have very different symbolical contents in different cultural contexts.

5 Feuer (1963) also gives a detailed account of the new emerging philosophy of power through science and technology. He claims that the new meaning of power is based not only on utilitarian values, but also on hedonist values.

6 Note the specific situation in the centre of France: one sector alone represents nearly 40 per cent of high altitude flights in Europe.

References

Balandier, G. (1985) *Le détour. Pouvoir et modernité*, Paris: Fayard.

Beniger J. (1986) *The Control Revolution*, Cambridge (MA): Harvard University Press.

Berthoud, G. and Busino, G. (eds) (1991) 'Les enjeux de la technique', focus issue, *Revue Européenne des Sciences Sociales* 29, 91.

Braun, I. and Joerges, B. (1994) *Technik ohne Grenzen*, Frankfurt am Main: Campus.

Bucholz, A. (1994) 'Armies, railroads and information: the birth of industrial mass warfare', in J. Summerton (ed.) *Changing Large Technical Systems*, Boulder (CO): Westview Press.

Caron, F. (1985) *Le résistible déclin des sociétés industrielles*, Paris: Perrin.

Daumas, M. (1991) *Le cheval de César, ou le mythe des révolutions techniques*, Paris: Editions des archives contemporaines.

Erhardt, J. (ed.) (1990) *Netzwerkphilosophie*, Hanover: Datacom.

Feenberg, A. (1995) *Alternative Modernity: The Technical Man in History and Philosophy*, Berkeley (CA.): UCLA Press.

Feuer, L.S. (1963) *The Scientific Intellectual: The Psychological and Social Origins of Modern Science*, New York: Basic Books.

Foucault, M. (1974) *Les Mots et les Choses, une archéologie des sciences humaines*, Paris: Gallimard (1st edn 1966).

Gras, A. (1979) 'Time of evolution and the spirit of time', *Diogène* 108: 57–84.

—— (1980) *Sociologie des ruptures, les pièges du temps en sciences sociales*, Paris: PUF.

—— (1997) *Les macro-systèmes techniques*, Paris: Gallimard (Que sais-je?).

—— (with S. Poirot-Delpech) (1993) *Grandeur et dépendance, sociologie des macro-systèmes techniques*, Paris: PUF.

Gras, A., Moricot C., Poirot-Delpech S. and Scardigli V. (1994) *Faced with Automation: The Pilot, the Controller and the Engineer*, Paris: Publications de la Sorbonne (French edition: *Le pilote, le contrôleur et l'automate*, Paris: ed. de l'IRIS, 1992).

Grundmann, R. (1994) 'Car traffic at the crossroads: new technologies for cars, traffic systems, and their interlockings', in J. Summerton (ed.) *Changing Large Technical Systems*, Boulder (CO): Westview Press.

Hughes, T.P. (1983) *Networks of Power, Electrification in Western Society 1880–1930*, Baltimore (MD): Johns Hopkins University Press.

Janicaud, D. (1995) *The Power of the Rational*, Princeton (NJ): Princeton University Press.

Juhlin, O. (1994) 'Information technology hits the automobile: rethinking road traffic as social interaction', in J. Summerton (ed.) *Changing Large Technical Systems*, Boulder (CO): Westview Press.

Kaijser, A. (1995) *I Fädrens Spår*, Stockholm: Carlsson Bokförlag.

La Porte, T.R. (1988) 'The United States air traffic system', in R. Mayntz and T.P. Hughes (eds) *The Development of Large Technical Systems*, Frankfurt am Main: Campus, and Boulder (CO): Westview Press.

Latouche, S. (1995) *La méga-machine*, Paris: La Découverte.

MacNeil, W.H. (1982) *The Pursuit of Power*, Chicago (IL): University of Chicago Press.

Maignan, G. (1991) *Le contrôle de la circulation aérienne*, Paris: Presses universitaires de France.

Mauss, M. (1936) 'Les techniques du corps', *Journal de Psychologie* 32: 271–93.

Offner, J.M. (ed.) (1994) *Réseaux, territoire et organisation sociale*, Paris: La Documentation Française.

Parrochia, D. (1993) *Philosophie des réseaux*, Paris: Presses universitaires de France.

Perrow, Ch. (1984) *Normal Accidents: Living with High Risk Technologies*, New York: Basic Books.

Perry Robinson, J.P. (1978) 'Neutron bomb and conventional weapons of mass destruction', *Bulletin of the Atomic Scientists* (March): 43–5.

Rochlin, G. (1996) *The Computer Trap*, Princeton (NJ): Princeton University Press.

Salomon, J. (1992) *Le Destin technologique*, Paris: Balland.

Scardigli, V. (1993) *Les Sens de la technique*, Paris: Presses Universitaires de France.

Studenyi, Ch. (1995) *L'Invention de la vitesse*, Paris: Gallimard.

Weingart, P. (1989) 'Grosstechnische Systeme: ein Paradigma der Verknüpfung von Technikentwicklung und Sozialen Wandel', in P. Weingart *Technik als soziale Prozess*, Frankfurt am Main: Suhrkamp.

Wise, J.A., Hopkins, V.D. and Smith, M.L. (1992) *Automation and Systems Issues in Air Traffic Control*, Berlin: Springer.

Part 4

CONCEPTUALIZING GOVERNANCE FOR INTERORGANIZATIONAL NETWORKS

12

ARE THERE SUCH THINGS AS SMALL NETWORKS?

Jean-Marc Offner

The current fascination for networks of every kind among our contemporaries needs some interpretation, as well as some perspective. The French notion of 'réseau' has a history, which is not the same as the history of 'network objects'. Sometimes behind, and sometimes ahead of, the development of infrastructure networks and 'networked organizations', the notion of network has had a relatively autonomous evolution.

The notion of large technical systems (LTSs) is more recent in origin and does not carry the intellectual and ideological inheritance which has conferred on the notion of network a problematic polysemy. Nevertheless, researchers reflecting upon the notion of LTS sometimes make use of the notion of network. Likewise, discourses on networks sometimes feed on the vernacular of systemics. An epistemological debate is then wide open: do large technical systems constitute an avatar, an up-to-date modality, of networks? Or are LTSs a particular socio-technical form?

To respond to this double question, we shall begin by highlighting the ambiguity of the notion of network – an ambiguity which paradoxically endows it with a certain efficacy. Then we shall present the history of the French notion of 'réseau' as a series of partly complementary paradigms. This will allow us to identify some of the recurring attributes of networks. In conclusion, we shall compare network attributes with LTS attributes. We shall emphasize in particular the role of territories in network growth in order to understand what, if anything, spatiality has to do with the 'large' dimension of large technical systems.

The notion of 'réseau': an ambiguous epistemological status

Has the French-speaking scientific community got sidetracked in endless speculations about networks, when elsewhere they are hardly discussed at all? Some have interpreted this specificity as an effect of the French language, the

French word 'réseau' being more polysemous than 'network' in English or 'Netz' in German (Dupuy 1988a). The doctrine of Saint-Simon, another French cultural peculiarity (see below), is also part of the story.

But, considering that social science discourses are becoming saturated with the notion of network, it is no longer possible at this point to do without at least minimal epistemological reflections concerning the term itself.

For some, there are networks . . . and there are networks. Bruno Latour, pondering over the choice of the word 'network' in light of his constructivist perspective on innovation, writes:

> If one means 'social network', misinterpretations are inevitable, for the networks of innovational sociologists mix many 'actants' together, very few of which have a human form. If one means 'technical network', the misunderstanding will not be any the less serious, since the arrangements in question are not objects as such, but projects, dreams, endeavours, or even entire societies.
>
> (Latour 1992: 2)

Thus it would seem that several notions of network coexist: the social networks of the sociologists, the technical networks of the engineers, the territorial networks of the geographers, the corporate networks of the economists and the 'technico-economic networks' of Latour and Callon. As Michel Callon once put it, 'given the number of uses [of the notion of network] that already exist, there is bound to be a misunderstanding!' (Callon 1992).

For others however, over and above diversity in reticularity ('networkness'), the question is one of 'trying to seize the unity of all networks in their differences, no matter what their substrata, their dimensions, their configurations (Parrochia 1993). For, 'what do a country, a body, a soul even, boil down to, if not to the superposition of a series of complex graphs (an entanglement of vertices and arcs)?' The network is thereafter used as the model which unifies: the world is a network, and we apprehend the universe as a network ('The human spirit can do only two things,' Bogdanov said, 'join together, or take apart').

The network is thus a monolithic, even monopolistic concept for one group, and a many-sided notion to be used differently according to discipline or theme for another. A more critical analysis explains the success and efficiency of the network discourse by the very ambivalence of the term: 'The network opens things up; the network closes things in. The whole symbolic power ("*imaginaire*") of the notion lies in this paradox' (Mercier 1988). The list of the antagonistic connotations of the notion of network is indeed quite long: ruses of the web vs control by hierarchies; central government vs self-management; public monopoly vs market instrument; rigidity of capital-intensive infrastructures vs organizational flexibility; and so on. Anne Cauquelin and Lucien Sfez go even further when the former evokes 'an archaic reference to the body' (Cauquelin 1988) when exploring the objects of modernity, while the latter notes that in this constant state of ambiguity, 'the network offers itself to all as a technology of the mind. One must think in terms of network, whether to camouflage a cold future or to a preserve a warmer past' (Sfez 1988). This confusion (agreed upon, consented to, or unconscious?) is all the more useful in that the network constitutes an exceptional instrument for representing reality – a representation which of course is partial, biased and impoverished, but which lets the world be seen simply, and which comforts the universalizing ambitions of the notion:

> The successful cultural integration of systemic analysis into our contemporary modes of thought undoubtedly derives in part from its simple ways (circles and arrows) of graphically representing complexity . . . The relationship between the notion of network and graphic representation seems to be even stronger. Using nodes and curved lines to draw infrastructures or organizations is symbolically, but also undoubtedly operationally, 'networking'.
>
> (Offner 1990: 61)

And, with this, a critical debate is opened: should the notion of network – overburdened as it is with meanings and metaphors, equivocal in its apparent simplicity – be unmasked now as the agent of a dominant ideology? Should it be pared down so that it may preserve a heuristic value? Or should more steady, basic notions be preferred instead – notions to be selected from among those already in contact with the notion of network? Finally, in the incessant comings and goings between the evolution of the world and the transformations of our representations of the world – Saussure explained that 'the vision constructs the object'; in the present case, we should specify: the 'focal length' of observation –, is the network located on the side of the object observed, or on that of the instrument of observation? To answer these questions, a detour into the history of the notion of network will be necessary. This history has in fact been well documented, but it is often given in fragments; we here propose a more coherent and reasoned view, using successive, cumulative more than concurrent, paradigms.

Two centuries of genealogy

Regulatory systems, hierarchy and web, territorial homogenization – History is not lacking in technical and organizational innovations containing the seeds of what would later actually be called 'networks'. Hence, the Romans possessed a sophisticated water distribution network, which followed a priority order for water supply in case of insufficient flow: first the public fountains, then the public baths and finally individual dwellings (Dupuy 1985). In fourteenth-century France, under Louis XI, the state began to run postal services which were administered in exactly the same way as our modern communications networks.

Several years before the French revolution, the streets in Paris were organized hierarchically, following a process for opening up and aligning the streets of the city (Darin 1986). It was during the same time that an engineer from the Ecole des Ponts et Chaussées, Achille-Nicolas Isnard, published (in 1781) his *Traité des richesses* ('Treaty on Wealth'), in which he sketched out 'a theory of locations, in which may be found his attempt to establish the characteristics of an optimal transportation network' (Lepetit 1989). In 1794, the Chappe telegraph system (optical semaphore) played a role in establishing national cohesion. The network spread rapidly, to the extent that several circuits (notion of connectivity) between Paris and the South of France were proposed, thus increasing the reliability of the communications.

So prior to the nineteenth century, networks and socio-technical systems functioning as networks were indeed in existence. But they were not conceptualized as such. Thus, until the beginning of the 1800s, French roads were not thought of in terms of network (one might also say, they did not constitute a 'system'). The absence of a national nomenclature, the scale of the projects (roads concerned surveyors more than geographers), procedures for financing, type of representation of local economic interests: all these factors account for the fact that the highways were at the time an 'unthinkable network' (Lepetit 1988a). In the same way, at the end of the eighteenth century, one still spoke of water (supply) or sewer 'lines' (Guillerme 1988).

The premise for the network existed. These may be seen in the evolution of concepts of the French military corps of engineers in defence strategies (Guillerme 1991); in the creation of new categories of thought among doctors, mechanical engineers, and mathematicians (Laterrasse and Coutard 1993). But it was the doctrine of Saint-Simon which elevated the network to the rank of a concept.

The network as a circulatory machine

Through an organicist metaphor, the doctrine of the Comte de Saint-Simon (1760–1825) – as transmitted and enriched by his numerous and influential epigones – synthesized two of the most powerful currents of nineteenth-century thought: the economics of circulation, and the utopia of the universal link (Mattelart 1994). On the one hand, the idea of the importance of traffic (Marx himself would speak of *Verkehr*), and of progress generated by a society in constant movement; the pre-eminence of the liquid over the solid state, of flows over stockpiles – ideas naturally linked to the 'invention' of the division of labour and free trade: for 'social physiology . . . money is to the political body what blood is to the human heart' (Saint-Simon). On the other hand, the desire for a world of solidarity, at peace, where science makes it possible to shift 'from governing men to dealing with things' (Saint-Simon); a socializing universalism which was the precursor of the urban planning and technocratic practices of the twentieth century.

The doctrine of Saint-Simon joined these two doctrines together in the notion of network: 'Everything may be put into relationship with everything else and generate a circulation of persons, goods, capital, ideas, knowledge' (Dupuy 1988a). Between the technical and the social realms, between infrastructure and organization, the network is both operational concept and social utopia, as these quotations from Michel Chevalier, one of the principal disciples of Saint-Simon, illustrate:

> Sophisticated means of transportation will reduce the distance not only from one point to another, but also from one [social] class to another [. . .] Industry consists of production centers connected by means of a relatively material link, namely, transportation routes, and by means of a relatively spiritual link, namely, banks . . . There is such a close relationship between the banking network and the transportation network, that when one of the two has been traced out in the most appropriate shape for the best exploitation of the planet, the other is to be be found similarly determined in its essential elements.
>
> (Chevalier 1832: 1)

Saint-Simonism was not just a kind of religious ideology; it also became a practice, in particular when it inspired the engineers who graduated from the Ecole polytechnique (Picon 1994) to construct railroads and canals, or to create – under the Second Empire – the large French financial institutions.

The railroad network was certainly the most concrete example of the doctrine of Saint-Simon. Whereas the original concept behind the railroad, in England as in France, was to provide an outlet for the industries situated within a particular locale (literally, to free industries from local isolation), the projects implemented by Saint-Simonian engineers aimed at the constitution of national, or even international communication systems (Ribeill 1990). The principle of network interconnection and the idea of a network of networks were already present: 'A reticulated society in which the methods for the circulation of knowledge, money, and means of communication' were interwoven (Musso 1988).

It is indeed difficult not to perceive the ideological similarities between the messianic discourse of the Saint-Simonians and the current declarations of the new prophets of networks, speaking of 'superconnectivity' (Hiltz and Turoff 1993) or predicting the advent in the third millennium of the 'symbiotic man', a 'being made of flesh and feeling, associated in a close symbiosis with a planetary organism which he helps to bring to life. A macro-organism constituted by the totality of men and their machines, nations and the large communication networks' (de Rosnay 1995).

The career of Saint-Simon's disciple Enfantin should also be recalled here. During the 1850s he worked on the fusion of the railroad companies, then on the creation of companies for urban transport and gas and water distribution. The mergers of large network operators recently brought about through deregulation in transportation, energy and telecommunications, are no surprise for historians of networks. The 'networked city', which has now been analysed (Tarr and Dupuy 1988, Dupuy 1991), has its roots in Haussmann's town planning, and was the application of Saint-Simon's precepts relating to the city: technical networks (water, gas, electricity), public transportation and a road network which allowed free circulation both within city limits and between cities.

Let us summarize this first, fundamental acceptance of the notion of network. The network as a circulatory machine offered a 'logistic' (the word entered military vocabulary in 1842) vision of society, in which means of production and of communication are totally linked together for the collective well-being. Flux and organic totality – the systemic discourse is not far off. Translated into contemporary terminology, Saint-Simon's proposals would not be disavowed by supporters of complexity: 'The more numerous, subtle and complex the internal relations of an organized body, the better this body is organized (passage from quantity to quality) and the more it produces an effect on its environment (passage from internal to external)' (Musso 1988).

The network as a relational graph of customers

At the end of the nineteenth century, a new meaning for the notion of network creeps in, concomitant with the development of electrical and telephone networks, which emphasizes the ideas of relations, web and service. For electricity, as for the telephone, the architecture (the configuration) of the network constitutes an economic factor of primary importance. There are network effects, related to the number and type of customers, to the size of the network and to the temporal and spatial diversity of uses.

For the railroad, these characteristics were not visible during the growth phase. The railroad companies appeared, certainly, as the first modern enterprises (Chandler 1977), but for quantitative rather than qualitative reasons (the size of the investments, in particular). If Chandler insisted on the theme of administrative coordination, it was to emphasize that the new practices and procedures invented by the managers of the railroad companies would ultimately be applied to all industrial enterprises. In contrast, in the case of electricity the history of the Chicago Edison Company clearly illustrates the specificity of certain management methods for network companies (Hughes 1983). The challenge of the 'universal system' was understood in the same terms, and at the same moment in history, by the managers of telecommunications systems.

This (historically) second meaning of network as a relational graph differs in two respects from the previous one (the network as a circulatory machine). First, it is concerned with corporate and not political economics; second, the final goal of the network manager is no longer to make fluxes circulate over a territory, but to ensure the technical and economic control of the network through a judicious set of relationships among customers. None the less, it is difficult not to observe the family relationship between the notion of organic totality on the one hand and universal service and monopoly on the other, inasmuch as they translate into operational practices. What had been a philosophical project for Saint-Simon (first meaning) now became an economic advantage for those who developed networks (second meaning).

Several decades later, graph theory – applied essentially to transportation networks (Kansky 1963, Garrison 1990) – and network economics (transportation, energy, telecommunications), formalized, elaborated and theorized this vision of the network as set up by the first managers of electricity and telephone networks. Network externalities, club effects, rate averagings: the vocabulary of the economists began to expand. Economists even provided definitions of networks: 'perhaps the best way to look at a network is as a cost sharing arrangement between several users' (Noam 1991); 'networks are complex technico-economic objects, the function of which is to put suppliers and customers for certain goods and services into a relationship with each other' (Curien 1992). The so-called three-layer theory clarifies the stratified structure of networks: infrastructures (support network), intermediary control services (command network) and ultimate consumer services (service network) (Curien and Gensollen 1991).

Let us summarize this meaning – the second chronologically – of the network: a relational graph of customers. The point of view here is microeconomic. Electricity and the telephone are the paradigmatic examples: meshing and universal service are key words. Here again, complexity and a systemic approach are in evidence.

The network as a spatio-temporal information switching device

Three American texts in the 1960s embodied a new vision of networks: *Order in Diversity: Community without Propinquity*, by Melvin Webber (1963); *Between Two Ages. America's Role in the Technetronic Era*, by Zbigniew Brzezinski (1969); and *War and Peace in the Global Village* by Marshall MacLuhan and Quentin Fiore (1969). We shall not analyse the contents of these books – which could be criticized for their technological determinism – but rather emphasize their historical context, during which the preceding representations of networks were abandoned in favour of the idea of a spatio-temporal information switching device.

In the majority of the developed countries during the 1960s, television and the telephone became common household items. The expression 'electronic communication' was getting popular; the airplane began to be a common means of transportation. In France this 'revolution' in communications took on a slightly different form. In fact, it was not until the middle of the 1970s that the French telephone network began rapid expansion; the Minitel and the high-speed train ('train à grande vitesse', or TGV) were innovations of the 1980s which contributed enormously to the development of this third paradigm of the network, centred around the notions of connectivity and speed. One started talking of ubiquity and challenging the idea that proximity and contiguity were the organizational principles of territories. The network became a tool for relating things to each other over time and space.

Of course, these ideas were not new. Already, at the time of the Chappe Brothers' telegraph, the near-instantaneousness of communication was admired. The 'transportation revolution' also dates from that crucial period at the turn of the eighteenth century. Speed, which had remained practically unchanged for several centuries, suddenly increased with the progress from the horse-drawn carriage, to the stagecoach, and then to the mail coach. The railroad fulfilled the need for rapid movement, which was already beginning to be felt on the roads: 'a journey from Paris to Marseilles required 359 hours of travel in 1650, 184 in 1782, 112 in 1814, 80 in 1834; in 1887 after the entire railroad track had been laid, the journey required less than a day, only 13 hours from one end to the other' (Studeny 1995). During this same period, Haussmann's Paris became a city of 'people in a hurry', where 'the straight, wide street with smooth traffic flow seems to lessen the distance and, as it were, put two points which before had seemed leagues apart to us in touch with each other', asserts the *Paris-Guide* written by Admond About in 1867 for the visitors to the Universal Exposition (Roncayolo 1983).

In spite of all this, there were still no intellectual tools that would help understand how communications networks shaped society – now seen as a 'relational planet' (Bressand and Distler 1995). By the end of the nineteenth century, the railroad and the stock market (whose history is closely linked to that of networks) had already granted time an economic value – minute by minute. But only the emergence of the notion of information, in the middle of the twentieth century, made it possible to recognize in networks the instruments, first, of urban structuration (Meier 1962), and then, of the 'globalization' of economic and social life:

> A technological revolution of historic proportions is transforming the fundamental dimensions of human life, namely time and space. New scientific discoveries and industrial innovations have extended the productive capacity of working hours while superseding spatial

> distance in all realms of social activity. The unfolding promise of information technology opens up unlimited horizons of creativity and communication.
>
> (Castells 1989)

The network as a relational graph of customers was the economists' notion. The network as a spatio-temporal information switching device belongs to geographers. This is because the exchange of information is involved in the existence of 'territorial nodosities' (Raffestin 1980) – singular points of a given territory which make relationships and circulation legitimate. 'At the outset, with the notion of network, the assertion of fundamental diversity and heterogeneity in time and space must be recognized' (Dupuy 1987).

Let us sum up this third notion of network as a spatio-temporal information switching device. The perspective is both topological and chronological: connectivity and speed. As with the Saint-Simonians, the notion implies circulation – now more immaterial than material. Through the notion of information, network and system are related again. Current discussions about the information superhighways are fully part of this paradigm of connectivity and speed. In fact they seem to echo Saint-Simonian discourses about political and economic progress of more than a century and a half ago.

The network as a decentralized coordinator

If we consider the use of the word 'network' in ordinary language (business 'networks', research 'networks', various social 'networks'), we are forced to admit that this is how the majority of people perceive networks: as a decentralized device for coordination. Once more, we shall refer to American writers who invented this 'organizational transaction' paradigm: Williamson (1981) with the theory of transaction cost; Granovetter (1985) and the notion of embeddedness.

Both authors belong to a current of economic sociology (Swedberg 1987). Their uses of the notion of network to understand economic behaviour are complementary. But for Williamson, the network represents essentially a coordination device located somewhere 'between market and hierarchy', in his words: it is a specific mode of governance of transactions. For Granovetter, networks of interpersonal relationships intervene directly in economic activity, and social networks construct economic institutions. Networks among institutions for one author, networks within organizations for the other; but, for both, the network does away with the opposition between micro and macro: individual behaviours produce organizations through successive local adjustments.

The sociology of innovation naturally appropriated this dual theory of the network. An emergent phenomenon par excellence, innovation constructs its own system of actors and transforms it. In the same way, the network procures the many relational contacts necessary to coordinate the process. Thus, for Michel Callon,

> a technico-economic network is a coordinated group of heterogeneous actors. [. . .] Technical-economic networks are composite: they mix the human and the non-human (technical devices . . .), [. . .] their dynamics may be understood only when related to the operation of translation which inscribes the inter-definition of the agents in the intermediaries put into circulation.
>
> (Callon 1991: 196)

In this fourth paradigm of the network, the nature of the relationships becomes a predominant element of analysis. These relationships are characterized as non-hierarchical, 'horizontal'; the network is meshed, not arborescent. The suppleness and flexibility of the network are emphasized when describing relations, the non-formal character of which is a guarantee of their dependability. Borrowing from the vocabulary of telecommunications, the specialists of social networks (Degenne and Forsé 1994) also speak of 'multiplexing' to designate relations which juxtapose several types of exchanges (personal and professional, for example).

Political science has also recently discovered the heuristic nature of the notion of network, to take into account the fragmentation of political power (Atkinson and Coleman 1992, Le Galès and Thatcher 1995). The network becomes a privileged mode of governance (Kooiman 1993). Although the definitions and uses differ according to researchers, the characteristics upon which everyone agrees are none the less recognizable:

> So far, we have a number of defining components, as well as dimensions, according to which networks may vary, but within a certain range only, in order not to lose their character as policy networks: being anchored in policy sectors; requiring collective action; composed of corporate actors; structured as interorganizational relations; predominantly informal and horizontal, but not without asymmetric interdependencies which means power relations; functionally defined by the formulation and implementation of policy: without stable central or hegemonic actors: involving not too many participants; and characterized by strategic interaction and a predominance of antagonistic cooperation or mixed-motive games.
>
> (Marin and Mayntz 1991: 18)

Interactive computer modes – Internet, to take one recent example – provide a powerful technical analogy for this paradigm of organizational transactions. As Michel Serres stated over 25 years ago, network and complexity share the same refusal of linearity: '[A network] is formed, for a given moment [. . .] of a plurality of vertices linked together by a plurality of arcs (paths) [. . .] By definition, no vertex is privileged in relation to others, and no vertex is univocally subordinated to another' (Serres 1969).

Need it be emphasized? The archetype of computers emerges from a social vision – 'a vision of the economics of social relationships . . . which derives from a cybernetic conception of human cooperation' (Pavé 1989). The analysis of concrete processes of information demonstrates perfectly the conditional aspect of the informational factor in organizational structuring, either centralizing or decentralizing according to circumstances.

Let us sum up this fourth and final meaning of the notion of network (final until when?). In the context of constructivist epistemologies (Le Moigne 1995), the network is a tool for the management (coordination) and comprehension (modelling) of complexity, in a world in which what joins together wins out over what takes apart and in which systems of action (economic, political, social) become contingent.

The notions of networks: unity and divergence

But perhaps this history of the notion of network (summarized in Table 12.1) is too French entirely. The division into phases here does not correspond to what Radkau had worked out (1991). The distinction made by Braun and Joerges (1994) between first- and second-order systems seems more in line with our typology, with second-order systems corresponding to the fourth paradigm – namely networks as decentralized coordinators – which no longer refers to technical networks, but to networked organizations. But, as we have

Table 12.1 Acceptations of the notion of network

Historic period	*Archetype*	*Representation*	*Paradigm*	*Privileged indicator*
First half of nineteenth century	Railroad	Circulatory machine	Flow and organic totality	Traces of infrastructure
End of nineteenth century	Electricity, telephone	Relational graph of customers	Networking and universal service	Architecture of the network
1960s	High speed transport, electronic communication	Spatio-temporal information switching device	Connectivity and speed	Accessibility
1980s	Conversational computers	Decentralized coordinator	Organizational transaction	Nature of the relationships

already made clear, it would be difficult to confuse the history of the notion of network with the history of network objects; the time spans are not the same. And, even if technical networks follow comparable cycles (Offner 1993), they do so at their own pace. Whatever the outcome of this debate, which is not crucial here, the chronological succession of four network paradigms raises questions about the unity of the concept.

First, let us point out that these paradigms all coexist today. The emphasis on large infrastructure networks in European construction was written directly into the doctrine of Saint-Simon. Debates over the past few years concerning new services (cable TV, mobile telephone, intelligent highways) tie up with the vision of the network as a relational graph of customers. The advent of multi-media is a continuation of the third paradigm (spatio-temporal information switching device). Finally, the network as a mechanism of decentralized coordination symbolizes successful management.

It should be noted next that paradigms succeed each other in relative harmony. From the circulation of flows to the meshes of the web, to speed, to transactions, the only difference is in the weight attributed respectively to the joint ideas of movement and relation. From organic totality to universal service, to connectivity, to organization, emerges over and over again the idea of totality.

Relation, totality: the kinship with the concept of system is obvious. Everyday language moreover, willingly substitutes one for the other: the railroad network or system, the telephone network or system, etc. To specialists of systems analysis (Walliser 1977), for every system there is a corresponding network of interactions between sub-systems. Thus, a technical network of communications allows for the functioning of a larger territorial system: city, region, country (Dupuy 1985). Nevertheless, the network – in its various meanings – contains specific connotations. Thus the network also helps to understand:

- heterogeneity, in geography as in sociology or in economics; the network permits an understanding of situations where competition and complementary co-exist;
- virtuality: whether technical or social, the network is a potential which needs to be activated (the potential of an address book, the potential of a transportation infrastructure on which different network-services may be exploited, etc.);
- redundancy: by means of the diversified circuits which it offers, thanks to meshing and connectivity, the network contains its own antidote for the risks of diffusion of incidents, from one point towards the entire network; dialectics of vulnerability and reliability.

Another feature which differentiates between network and system may be added to this list: the system presupposes a boundary, an 'inside' (the system itself) and an 'outside' (the environment). The situation is not so clear with networks (see Grundmann's contribution to this volume), and here we come up against the intrinsic duality of the notion. In the constructivist paradigm – the fourth one – the network is structured by actors, those 'little indefatigable workers' building their own action system with fluctuating limits and a specific scope. In other meanings of the notion of network, 'transactional projects' (Dupuy 1987) between actors are taken over by a 'network operator' (the notion is close to that of Thomas Hughes's 'system-builder'), inspired by a dual ambition: to create and preserve monopoly and to centralize.

On the one hand a network which may be termed self-regulating, in which local regulations add up to produce overall coherence (the invisible hand of the network . . .). On the other hand, the necessity for some organ external to the network to ensure management and development; economic requirements, political requirements:

> Here and there, the reiterated proof of a fundamental aversion to the notion of network crystallizes: a refusal of the principles which lead to the setting up of reticulated configurations, a will to persist in controlling the network – even when it has become impossible – recalling a time when direct control was still possible.
>
> (Stourdzé 1987: 135)

The changes taking place with the Internet are a fine example of the transition from one mode of network operation to another.

Let us try to conclude this brief discussion of the attributes of the notion of network, based on the study of its genealogy. The core of the notion is centred around the ideas of relation and totality; and, in this, network and system are close neighbours. But the notion turns out to be ambivalent when one discusses network regulation and the management of network effects: self-management through local vs centralized regulatory control. This dichotomy – which does not in any way correspond to the distinction between technical and immaterial networks – will be important in what follows, if it is found to play a role in the 'size' of the network, and thus in the conceptual relationships between the network and the large technical system.

Networks and large technical systems: LTSs in search of the spatial

The notion of 'large technical system' has one very large advantage over that of network: 'LTS' is not a household term. Still, the semantic controversies surrounding the term are not any the less numerous. Starting with the definitions initially suggested by Hughes (1983) and Joerges (1988), everyone agrees about a unitary substratum, where the notions of system (naturally), but also heterogeneity, coupling, complexity and integration are to be found (Gras 1990). The next step is to track down the clues which elevate a research object to the rank of an LTS. One may ask, for example, whether road traffic is an LTS (Grundmann 1994).

Obviously, the notional fields of network and LTS overlap. Most of the case studies carried out by researchers of the LTS 'school' concern technical networks. Joerges (1988) classifies 'Large Technical Networks' as a variation of the LTS, but is in this case referring rather to interorganizational socio-political networks. Alain Gras, referring specifically to the development of the railroads, electricity and telephone, emphasizes that 'the network is deeply rooted in the reality of the system' (Gras 1993). Furthermore, he claims that macro-systems are all communications systems, that their 'spatio-temporal fabric' assumes the form of a network.

In contrast, Hughes, like Gras – and Dupuy by another approach – retains the opposition between the network as the internal structure, or framework, and the system as something which opens into the environment. Is this a purely academic debate? Perhaps, if we consider the operating results of the use of the two notions. None the less, calling the telephone, or the high speed train, a 'network' makes it possible to formalize as systems the territories or socio-economic activities to whose functioning these networks contribute. Calling the same telephone or the same high speed train a 'system' requires that one specifies what one takes away from the system (telephone, train) so that the associated networks may appear – but then the nature of these networks seems unclear.

Only the question of the scale, or the size of the technical system, will make it possible to advance this debate. We have already suggested above that the notion of network is ambivalent with respect to two possible forms of regulation; these modes of dealing with network effects are perhaps linked to size.

For Hughes, LTSs are bound to grow: the system spreads out to become more efficient, optimizing the interactions between supply and demand by the increase of the diversity of demand ('load diversity') and supply ('economic mix'). Kornwachs (1993) explains this process with remarkable concision: LTSs can be managed only through growth. The evolutionist point of view predominates among LTS specialists, behind the figure of the octopus or the spider. LTSs are first local, then regional, then national, international and finally intercontinental. LTSs ignore borders, owing to the interplay of interconnection. And this causes headaches for researchers interested in 'small' technical networks. 'District heating systems are distinctly regional or local systems. Can district heating systems really be regarded as large sociotechnical systems?' (Summerton 1992). In the name of relativity, the urban heating system of Mjölby will be given an LTS label!

The thesis of the irreducible enormity of LTSs seems to us to pay little attention to both empirical observations and theoretical teachings. As

Braudel has admirably demonstrated for the French case, one should first place the emphasis on the role of networks as factors of national cohesion. This was true in the nineteenth century, and this still in large part holds true for the twentieth (see the French debate on the nation's territorial development). Unless it is held that politics is an external variable in the 'real world' of economics, one cannot ignore this reference to the network as the unifying instrument within a specific territory. Moreover, historical analysis weakens the thesis of a general evolution of LTSs from local to world-wide: 'initial conceptions and spatial configurations of networks are specific to each network. This differentiation concerns first the degree of integration with the networks of neighbouring territories. Certain networks were, from the outset, by reason of the nature of the flow they transported, conceived on an international scale' (Caron 1995). Finally, there are very real 'small' technical systems which possess all the characteristics of the large ones, size excepted. Such is the case for a good number of urban technical networks, which have no reason to 'de-urbanize'.

Indeed, it should not be forgotten that many collective services and utilities offer an alternative between point distribution or network distribution. One may buy mineral water at the store, or wait for the construction of a specific network of 'very good' drinking water; one may be connected to the sanitation network, or have one's own septic tank (especially in the country); and so on. One might retort that point distribution is a form of second-level LTS, which utilizes pre-existing technical systems. But this kind of argument would lead us in the end to decreeing that anything may be called an LTS – or network – including pot roast, the favourite example of debaters in the French research group on Networks (Amar 1993, Curien 1993).

On the theoretical level, it seems to us that there is a dual confusion here: between large size ('large scale') and territorial monopoly; between the idea of integration and that of universalization.

The now well-documented history of the forms of management of technical networks (Simon 1991, Stoffaës 1994) emphasizes the role of the monopoly principle within territories defined either economically (*functional* territories) or politically (*institutional* territories). The legitimacy of the monopoly derives from network growth, but this growth is limited to the monopoly zone. We shall use the words 'intensive growth' to designate the development of the web inside a territory and the possible integration of specialized networks into one single network through the homogenization of flows (different electric currents; different types of information, etc.). In this first type of growth there is no modification of scale.

Other networks undergo a type of growth by means of interconnection; we shall speak here of 'extensive growth'. Heterogeneous networks – with different technologies, or management, or institutional features – are joined together to form a hybrid network (Dupuy *et al.* 1990), which is not just the juxtaposition of two initial networks made compatible through gateway technology (David and Bunn 1991), but a new network entirely, with partially specific characteristics. In this second type of growth, the intensification of the web is not the object, since the notion of universal service is absent (for example: express delivery services, or high-speed communication networks). Contrary to intensive growth, extensive growth modifies the scale of the network; but not always in order to enlarge it. For example, the creation of the Paris area rapid transit system ('réseau express régional', or RER), by the interconnection of the RATP (the Paris public transport system) and the SNCF (the French nationwide train carrier) networks created a new territorial level of service, a regional scale situated between the local and national scales of the pre-existing networks.

Distinguishing between the two types of network extension thus suggests that there is no correlation between growth and large scale. There exist forms of network growth which do not increase scale. Another way to question the thesis of an inevitably 'big' LTS/network is to point out the limits of the growth factors of the networks. To simplify, these are of two types: effects of range or diversity, and club effects (if we admit that economies of scale are not specific to networks).

The externalities of scope explain the extension of electrical networks (in order to take advantage, in particular, of the temporal diversity of consumption and production costs), and of water distribution networks (in order to increase service reliability). These externalities are limited by transportation costs, mainly due to power losses on electrical lines, and to infrastructure costs for water. In certain cases, the optimal size of the network will perhaps be the entire planet, in other cases it will be limited to a smaller area.

'Club' externalities (Curien and Gensollen 1992) – the fact that each new customer on a network benefits from the presence of previous customers – explain the growth of communications, transportation or telecommunications networks. Even so, there has to be a socio-economic or socio-political relevance for the new members of the 'club' with respect to the old members. Various projects for trans-European infrastructures were justified by the creation of the Common Market; Internet by the existence of an international scientific community. As has already been clearly demonstrated with respect to the emergence of national train systems (Salsbury 1995) networks and territories are congruent. Here again, there are cases in which the appropriate territory will be the entire planet, others a more limited space. The present-day internationalization of networks is thus not due to a specificity of networks, but rather to the internationalization of society, which gives meaning to planetary networks.

The two major growth factors are thus both limited. Whether we address the question of the size of networks in terms of how they grow or of why they grow, the idea of a continuous extension of networks appears unfounded. It seems to us that similar reasoning holds for LTSs.

The topic of the regulatory mode of networks, in contrast, has not appeared as a discriminating variable for thinking about the size of networks. 'Centralized' networks may be small, 'autonomous' networks may be large, and vice versa. The analysis of the effects of deregulation on network structures should confirm this opinion.

Directions for research

The territorial perspective on the notion of network – developed in particular in the French CNRS Research Group on Networks (Groupement de recherche 'Réseaux') – undoubtedly challenges the notion of 'large technical system', unless 'large' refers to a non-spatial dimension of the system. If 'large' refers to specific modes of development, as was argued in this chapter, then relations can be established between the notions of network and that of LTS, as mapped by Joerges (1988) and Braun and Joerges (1994) (see Table 12.2). The first three network paradigms correspond to the notion of (first-order) LTSs, and the fourth paradigm (the network as a decentralized coordinator) to second-order LTSs, which make use of first-order ones to organize and function. Thus a blood collection network, a 'city network', or firm (second-order) networks will use transportation, telecommunications and information (first-order) networks.

This raises once again the questions of how to control these networks (systems), and of which indicators are relevant to qualify their various forms. We argued here that first-order LTSs are centrally controlled, whereas second-order ones function in a decentralized way, through the combination of local arrangements. In order to validate this assumption, it would be necessary not only to describe accurately the many tasks that control consists of, and the relations between these tasks; it would also require the empirical study of the evolution in the modes of control of centralized systems currently subject to

Table 12.2 Relations between the notions of network and LTS

Network		*LTS*
Circulatory machine Relational graph of customers Spatio-temporal information switching device	'Centralized' networks	First-order LTS*
Decentralized coordinator	'Autonomous' networks	Second-order LTSs** Large technical networks*

Notes

* See Joerges 1988.

** See Braun and Joerges 1994.

'deregulation'. Is control decentralized due to deregulation, or do new forms of centralized control emerge? This in turn raises the crucial question, scarcely addressed yet, of the organizational and institutional features of the 'network operator', the 'networker' – this socio-technical system consisting of actors, rules, norms, and various other devices, which is in charge of 'networking processes'.

References

Amar, Georges (1993) 'Qu'est-ce qui n'est pas un réseau?', *Flux, Cahiers scientifiques internationaux Réseaux et Territoires* 13/14: 56–8.

Atkinson, Michael M. and Coleman, William D. (1992) 'Policy networks, policy communities and the problems of governance', *Governance* 5, 2 (April).

Braun, Ingo and Joerges, Bernward (1994) 'How to recombine large technical systems: the case of European organ transplantation', in Jane Summerton (ed.) *Changing Large Technical Systems*, Boulder (CO): Westview.

—— (eds) (1994) *Technik ohne Grenzen*, Frankfurt am Main: Suhrkamp.

Bressand, Albert and Distler, Catherine (1995) *La Planète relationnelle*, Paris: Flammarion.

Breton, Thierry (1991) *La dimension invisible, le défi du temps et de l'information*, Paris: Editions Odile Jacob.

Brzezinski, Zbigniew (1969) *Between Two Ages: America's Role in the Technetronic Era*, New York: Viking Press.

Callon, Michel (ed.) (1989) *La science et ses réseaux*, Paris: La Découverte.

—— (1991) 'Réseaux technico-économiques et irréversibilité', in R. Boyer (ed.) *Figures de l'irréversibilité en économie*, Paris: Editions de l'EHESS, 195–230.

—— (1992) 'Sociologie des sciences et économie du changement technique: l'irrésistible montée des réseaux technico-économiques', in Collection du Centre de Sociologie de l'Innovation (CSI), *Ces réseaux que la raison ignore*, Paris: L'Harmattan, 53–78.

Caron, François (1995) 'Intégration des réseaux et technologie', in Michèle Merger, Albert Carreras and Andrea Giuntini (eds) *Les réseaux européens trans-nationaux XIXe-XXe siècle, quels enjeux?*, Nantes: Ouest éditions, 403–8.

Castells, Manuel (1989) *The Informational City: Information Technology, Economic Restructuring and the Urban Regional Process*, Oxford: Blackwell.

Cauquelin, Anne (1988) 'Concept pour un passage', *Quaderni* 3 (Winter): 31–40.

Chandler, Alfred D. (1977) *The Visible Hand. The Managerial Revolution in American Business*, Cambridge (MA): Harvard University Press.

—— (1979) 'The railroad, a cross-cultural comparison', Colloque Science et décision (CNRS–MIT), Paris.

Charléty, Sébastien (1964) *Histoire du Saint-Simonisme*, Paris: Gonthier.

Chevalier, Michel (1832) 'Système de la méditerranée', *Le globe, Journal de la religion saint-simonienne*, 12 February.

Commissariat Général du Plan (1995) *Les réseaux de services publics, dans le monde: organisation, régulation, concurrence*, Paris: éditions Aspe-Europe.

Curien, Nicolas (ed.) (1992) *Economie et management des entreprises de réseau*, Paris: Economica.

—— (1993) 'Réseau: du mot au concept', *Flux, Cahiers scientifiques internationaux Réseaux et Territoires* 13–14 (July–December): 52–5.

Curien, Nicolas and Gensollen, Michel (1991) 'Infrastructures, contrôle-commande, services: le hard et le soft', *Annales des Mines (series Réalités industrielles)*, April: 98–103.

Curien, Nicolas et Gensollen, Michel (1992) *Economie des télécommunications. Ouverture et réglementation*, Paris: Economica.

Darin, Michaël (1986) *Alignement des rues*, Paris: Bureau de la recherche architecturale, Nantes: Ecole d'architecture de Nantes.

David Paul A. and Bunn, Julie Ann (1991) 'L'économie des passerelles technologiques et l'évolution des réseaux', *Flux, Cahiers scientifiques internationaux Réseaux et Territoires* 4 (April–June): 15–29.

de Rosnay, Joël (1995) *L'homme symbiotique*, Paris: Seuil.

Degenne, Alain and Forsé, Michel (1994) *Les réseaux sociaux*, Paris: Armand Colin.

Dupuy, Gabriel (1985) *Systèmes, réseaux et territoires*, Paris: Presses de l'Ecole Nationale des Ponts et Chaussées.

—— (1987) 'Les réseaux techniques sont-ils des réseaux territoriaux?', *L'espace géographique* 3: 175–84.

—— (1988a) 'Réseaux', *Encyclopaedia Universalis*, Paris: 875–82.

——- (ed.) (1988b) *Réseaux territoriaux*, Caen: Paradigme.

—— (1991) *L'urbanisme des réseaux*, Paris: Armand Colin.

Dupuy, Gabriel, Gély, Corinne and Offner, Jean-Marc (1990) 'RER et interconnexions: les vertus d'un réseau hybride', *Flux, Cahiers scientifiques internationaux Réseaux et Territoires* 2 (Fall): 81–93.

Flichy, Patrice (1991) *Une histoire de la communication moderne*, Paris: La Découverte.

Garrison, William L. (1990) 'Networks: reminiscence and lessons', *Flux, Cahiers scientifiques internationaux Réseaux et Territoires* 1 (Spring): 5–12.

Granovetter, Mark (1985) 'Economic action and social structure: the problem of embeddedness', *American Journal of Sociology* 91, 3: 481–510.

—— (1992) 'Economic institutions as social constructions: a framework of analysis', *Acta Sociologica* 35, 1.

Gras, Alain (1990) 'La notion de macro-système-technique', *Culture Technique* 21 (July): 76–7.

—— (1993) *Grandeur et dépendance. Sociologie des macro-systèmes techniques*, Paris: Presses universitaires de France.

—— (1994) 'Pour une technologie pacifiée', *Le Monde* (11 January).

Grundmann, Reiner (1994) 'Car traffic at the crossroads: new technologies for cars, traffic systems, and their interlocking', *Flux, Cahiers scientifiques internationaux Réseaux et Territoires* 17 (July–September): 19–32.

Guillerme, André (1988) 'L'émergence du concept de réseau 1820–1830', in Gabriel Dupuy (ed.) *Réseaux territoriaux*, Caen: Paradigme, 33–50.

—— (1991) 'Réseau: genèse d'une catégorie dans la pensée de l'ingénieur sous la Restauration', *Flux, Cahiers scientifiques internationaux Réseaux et Territoires* 6 (October–December): 5–17.

Hiltz Starr, Roxanne and Turoff, Murray (1993) *The Network Nation* (revised edition), Cambridge (MA): MIT Press.

Hughes, Thomas (1983) *Networks of Power. Electrification in Western Society 1880–1930*, Baltimore (MD): Johns Hopkins University Press.

Joerges, Bernward (1988) 'Large technical systems: concepts and issues', in Renate Mayntz and Thomas Hughes (eds) *The Development of Large Technical Systems*, Frankfurt am Main: Campus.

—— (1995) 'Technique sans frontières', *Flux, Cahiers scientifiques internationaux Réseaux et Territoires* 20 (April–June): 40–5.

Kansky, Karel J. (1963) *Structure of Transportation Networks*, Chicago (IL): University of Chicago Press.

Kooiman, Jan (ed.) (1993) *Modern Governance. New Government–Society Interactions*, London: Sage.

Kornwachs, Klaus (1993) *Steuerung und Wachstum. Ein Systemtheoretischer Blick auf Große Technische Systeme*, Wissenschaftszentrum Berlin and FSII.

Laporte Todd R. (ed.) (1991) *Social Responses to Large Technical Systems: Control or Anticipation*, Dordrecht (NL): Kluwer Academic Publishers.

Laterrasse, Jean and Coutard, Olivier (1993) 'Concept de réseau et modélisation des systèmes complexes: quelques éléments pour une réflexion problématique', Second European Congress of Systems Theory (Prague, 5–8 October).

Latour, Bruno (1992) 'Introduction', in Collection du Centre de Sociologie de l'Innovation (CSI), *Ces réseaux que la raison ignore*, Paris: L'Harmattan, 1–4.

Lazega, Emmanuel (1994) 'Analyse de réseaux et sociologie des organisations', *Revue française de sociologie* 35: 293–320.

Le Galès, Patrick and Thatcher, Mark (1995) *Les réseaux de politique publique. Débats autour des 'policy networks'*, Paris: L'Harmattan.

Le Moigne, Jean-Louis (1995) *Les épistémologies constructivistes*, Paris: Presses universitaires de France.

Lepetit, Bernard (1988a) 'L'impensable réseau: les routes françaises avant les chemins de fer', in Gabriel Dupuy (ed.) *Réseaux territoriaux*, Caen: Paradigme, 21–32.

—— (1988b) *Les villes dans la France moderne (1740–1840)*, Paris: Albin Michel.

—— (1989) 'Présentation du texte d'A.N. Isnard: "De l'utilité des routes et des canaux navigables"', *Flux, Cahiers scientifiques internationaux Réseaux et Territoires* (special issue): 5–7.

Lévy, Jacques (1994) *L'espace légitime. Sur la dimension géographique de la formation politique*, Paris: Presse de la Fondation Nationale des Sciences Politiques.

MacLuhan, Marshall and Fiore, Quentin (1969) *War and Peace in the Global Village*, New York: Bantam.

Marin, Bernd and Mayntz, Renate (eds) (1991) *Policy Networks: Empirical Evidence and Theoretical Considerations*, Frankfurt am Main: Campus.

Mattelart, Armand (1994) *L'invention de la communication*, Paris: La Découverte.

Mattelart, Armand and Mattelart, Michèle (1995) *Histoire des théories de la communication*, Paris: La Découverte.

Meier, Richard L. (1962) *A Communication Theory of Urban Growth*, Cambridge (MA): MIT Press.

Mercier, Pierre-Alain (1988) 'La maille et le réseau', *Quaderni* 3: 41–9.

Merger, Michèle (ed.) (1995) *Les réseaux européens transnationaux, XIXe–XXe siècles, quels enjeux?*, Nantes: Ouest éditions.

Musso, Pierre (1988) 'Aux origines du concept moderne: l'organisme et le réseau dans la physiologie de Saint-Simon', *Quaderni* 3: 11–29.

Noam, Eli M. (1991) 'Network tipping and the tragedy of the common network: a

theory for the formation and breakdown of public telecommunications systems', *Communications et stratégies* 1.
—— (1994) 'Beyond liberalization: from the network of networks to the system of systems', *Telecommunications Policy* 18, 4: 286–94.
Offner, Jean-Marc (1979) 'Réflexions sur les origines du chemin de fer en France. Le contexte institutionnel, économique et culturel', Colloque science et décision (CNRS–MIT).
—— (1990) 'La représentation graphique des réseaux: outil heuristique ou innovation de gestion?', *Flux, Cahiers scientifiques internationaux Réseaux et Territoires* 2 (Fall): 61–5.
—— (1993) 'Le développement des réseaux techniques. Un modèle générique', *Flux, Cahiers scientifiques internationaux Réseaux et Territoires* 13–14 (July–December): 11–18.
—— (1994) 'Réseaux, territoires et organisation sociale', *Problèmes politiques et sociaux* 740 (December), Paris: La Documentation française.
Parrochia, Daniel (1993) *Philosophie des réseaux*, Paris: Presses universitaires de France.
Pavé, Francis (1989) *L'illusion informaticienne*, Paris: L'Harmattan.
Picon, Antoine (1994) 'Les polytechniciens saint-simoniens au XIXe siècle', *Notes de la fondation Saint-Simon* (July).
Radkau, J. (1991) *Zum Ewigen Wachstum Verdammt?, Historisches Über Jugend und Alter Großer Technischer Systeme*, Wissenschaftszentrum Berlin and FSII.
Raffestin, Claude (1980) *Pour une géographie du pouvoir*, Paris: Litec.
Ribeill, Georges (1990) 'Des Saint-Simoniens à Léon Lalanne: projets, thèses et controverses à propos de l'organisation des réseaux ferroviaires', *Revue d'histoire des chemins de fer* 2 (Spring): 47–80.
Roncayolo, Marcel (1983) 'La production de la ville', in Georges Duby (ed.) *Histoire de la France urbaine, la ville à l'âge industriel*, Paris: Seuil.
Salsbury, Stephen (1995) 'L'espace comme facteur de développement des réseaux à grande échelle', in Michèle Merger, Albert Carreras and Andrea Giuntini (eds) *Les réseaux européens trans-nationaux XIXème–XXème siècle, quels enjeux?*, Nantes: Ouest éditions, 397–402.
Serres, Michel (1969) *Hermès I, La communication*, Paris: Editions de Minuit.
Sfez, Lucien (1988) *Critique de la communication*, Paris: Seuil.
Simon, Jean Paul (1991) *L'esprit des règles. Réseaux et réglementation aux Etats-Unis*, Paris: L'Harmattan.
Stoffaës, Christian (ed.) (1994) *L'Europe à l'épreuve de l'intérêt général*, Paris: ASPE Europe.
—— (1995) *Services publics, question d'avenir*, Paris: Editions Odile Jacob.
Stourdzé, Yves (1987) *Pour une poignée d'électrons: pouvoir et communication*, Paris: Fayard.
Studeny, Christophe (1995) *L'invention de la vitesse, France, XVIIIe–XXe siècle*, Paris: Gallimard.
Summerton, Jane (1992) *District heating comes to Town. The Social Shaping of an Energy System*, Linköping (SE): Linköping Studies in Art and Science.
—— (ed.) (1994) *Changing Large Technical Systems*, Boulder (CO): Westview.
Swedberg, Richard (1987) *Current Sociology*, London: Sage.
Tarr, Joel and Dupuy, Gabriel (eds) (1988) *Technology and the Rise of the Networked City in Europe and America*, Philadelphia (PA): Temple University Press.

Virilio, Paul (1990) *L'inertie polaire*, Paris: Christian Bourgeois.
—— (1991) 'For a geography of trajectories', *Flux, Cahiers scientifiques internationaux Réseaux et Territoires* 5 (July–September): 48–54.
Walliser, Bernard (1977) *Systèmes et modèles*, Paris: Seuil.
Webber, Melvin (1970) (first pub. 1963) 'Order in diversity: community without propinquity', in H.M. Proshansky, W.H. Illetson and L.G. Rivlin (eds) *Environmental Psychology*, New York: Holt, Rinehart and Winston.
Williamson, Oliver E. (1981) 'The economics of organization: the transaction cost approach', *American Journal of Sociology* 87: 3.

13

ON CONTROL AND SHIFTING BOUNDARIES

Modern society in the web of systems and networks

Reiner Grundmann

Introduction

This chapter examines the relations between two basic concepts in social sciences, the concepts of 'system' and 'network'. It discusses the normative overtones (either manifest or latent) expressed by these concepts, and how the use of the concepts by different authors or in different approaches coheres with these overtones.

It is commonly held that systems approaches deal primarily with questions of control and regulation, whereas network approaches are better adapted to the study of loose structures. It turns out, however, that this is only true with respect to dictionary definitions. Recent theoretical developments in the field of social theory have lead to a paradoxical picture. Current system theories, it will be argued, avoid the notion of control altogether by stressing the self-regulating and non-hierarchical features of social systems, while current network theories seem to acknowledge the need for the maintenance and control of the social fabric.

The argument will be developed in three steps. I shall first discuss the 'conventional wisdom' on systems and networks and their normative properties, based on definitions drawn from social science dictionaries. Then I shall emphasize the contrast between these definitions and the use of the notions in the three contemporary social theories: Luhmann's social systems theory (SST), the large technical systems (LTS) approach, and actor-network theory (ANT). All three approaches take on the task of providing theoretical insights on social organization based on the notions of system and network, although they depict different basic structural features of modern societies. I will do this by focusing on the issues of control and boundaries. Finally, I shall analyse the significance of this contrast between the commonly accepted definitions of the two concepts and the ways in which they are used.

The notion of systems in the social sciences

In the *Critical Dictionary of Sociology*, Boudon and Bourricaud suggest that the term 'system' has a longer tradition and deeper foundation within the social sciences than the term 'network'. They assimilate Bertalanffy's definition of a system with that of Condillac.[1] At present, the standard reference for the introduction of the concept of network in social sciences is Radcliffe-Brown (1940), whereas the term system is associated with Bertalanffy (1956) and his general system theory.

Although introduced later, the concept of system has been widespread for a long time. Probably this has to do with its enthusiastic reception by mainstream (structural-functional) sociology and cybernetics from the 1940s onwards. It is only recently that the notion of network has become more common. This is probably due primarily to the development of social network analysis (as a highly formalized technique) which has become a genuine growth industry over the past decade or so.

I shall now briefly review definitions in social science dictionaries and analyse what they tell us about the properties of systems and networks, and how these properties influenced the reception of the two concepts.

The *Critical Dictionary of Sociology* quotes Bertalanffy's definition of a system: 'A system is an ensemble of interdependent elements in which *the elements and the whole system will change as a result of the change of one of them*' (my emphasis). The *Blackwell Dictionary of Twentieth Century Social Thought* introduces an additional criterion: 'Systems are complexes of elements and relations, separated by boundaries from their environment, which is always more complex than the system itself.' The *Social Science Encyclopedia* quotes *Webster's*: a system is a 'set or arrangement of things so related or connected as to form a unity or organic whole'.

Systems are thus defined as sets of things and their relations. The influence of mathematics is clearly visible, and Anatol Rapoport in the *International Encyclopedia of the Social Sciences* celebrates exactly this point when stating:

> Here . . . is a unifying principle, which truly abstracts from the content of phenomena and concentrates on the structural and dynamic relations . . . If we follow the definition of 'system' . . . as a specified set of entities and a set of relations among them – then it would seem that the method of mathematical homology is the most natural foundation of a general systems theory.

Issues of control are emphasized, especially in the closely related field of cybernetics, in connection with issues of accuracy. Ashby (1956) states that cybernetics analyses all forms of behaviour which are organized, determined or reproducible. The 'art of steering', as introduced by Wiener (1948), was initially applied to an anti-aircraft weapons system in which the built-in feedback mechanism guided missiles to their target. In order to be effective, the system had to be designed in a very precise manner. The slightest error in the 'initial conditions' of such a system would have made it useless and allowed the enemy (Japanese aircraft in that instance) to enter the country's air space.[2] Cybernetic systems have to be precise, because their output can only be ascertained if the input–output relation is clear and both the input signal and output operation are very precise.

As a reaction to the importance of control in the systems approach, interest in the concept of network arose in the 1960s, when the active dimension of human interaction was (re)discovered:

> The enthusiasm for network analysis was related to the theoretical shift away from structural-functional analysis which by 1960 had dominated social science in Britain and the United States for 30 years. The network concept permitted the entry on a systematic basis of interacting people engaged in actions that could *alter the institutions* in which they participated.
>
> (*Social Science Encyclopedia*: 'network', my emphasis; see also Homans 1964)

The only other dictionary under review which mentions networks, the German *Grundbegriffe der Soziologie*, defines them as loose couplings or clusterings of 'elements' (such as groups, organizations or institutions). The *Social Science Encyclopedia* claims that network analysis

> complements other research techniques in a number of ways: it focuses systematically on interlinkages and interdependencies between units of analysis. This focus embraces micro and macro levels in one analytical framework. It provides a systematic framework for analyzing tension and asymmetry in social relations and consequently highlights their inherent dynamics.

Approaches in terms of networks opened up a new area of research. The growing literature in the field testifies to the fruitfulness of the concept, especially for policy analysis (among others, see Heclo and Wildavsky 1974, Heclo 1978, Granovetter 1982, Jordan 1990, Kenis and Schneider 1991, Laumann and Knoke 1987, Marin and Mayntz 1991, Powell 1990).

The term network has affinities to such notions as fabric, web, knitting. This is in contrast to system, which has no direct metaphorical appeal. The semantic divide between system and network thus appears relatively clearcut. A review of some recent developments in social theory, however, suggests a more complex, somewhat paradoxical picture. I will examine this by focusing on two main aspects: control and boundaries.

Control

In this section, I shall discuss in turn how control issues are conceptualized in Luhmann's social systems theory, in the large technical systems approach, and in actor-network theory. In the next section, I shall turn to the question of boundaries.

Social systems theory

> Mit Worten läßt sich trefflich streiten, mit Worten ein System bereiten . . .
>
> Goethe

Niklas Luhmann analyses societies in terms of 'social systems'. Coming from a Parsonian background, Luhmann conceives of modern society as falling into specific subsystems which are functionally differentiated, yet interdependent. Science, law, politics and economics are the most prominent social systems in his approach. According to Luhmann, social systems are meaning-processing systems that deal with events (communications) and form structures, codes and programmes. Thus the basic elements of all social systems are communications, themselves produced by communications: social systems are *autopoietic* systems.[3]

Social systems are self-referential systems: modern society does not have a centre nor a 'top'. It is functionally differentiated into several self-organizing functional systems (see Figure 13.1). The intervention of one system in other systems in order to control, regulate or command is no longer possible (see Willke 1987 and the exchange between Luhmann 1989b and Scharpf 1989). In fact the control issue is absent from Luhmann's theory altogether. As Alan Wolfe recently observed, theories of self-organization are appealing because 'automatic, self-regulating systems are non-hierarchical, thereby overlapping with a political commitment to egalitarianism' (Wolfe 1993: 131).

Technical systems were initially not explicitly integrated in this approach, although in *Social Systems* Luhmann does give a mention to machines as systems (Luhmann 1995: 2). For Luhmann,

> The concepts of technology, technical, and technicalization shall denote the fact that operations are possible without too much reflection on the part of the subject or observer. In particular, there is no need for inquiries by subjects or observers. In this sense technology means a process of 'disburdening'.
>
> (Luhmann 1990: 197, my translation; see also Luhmann 1993b)

Technology reduces complexity; it also transforms uncertainty into secure expectations. In this respect, it differs from the rest of the environment of social systems which, Luhmann argues, provides these systems not only with resources, but above all with 'perturbation, disturbance and noise'.

This specificity of technology in Luhmann's theory can be explained. Luhmann's basic definitions of technology either as a 'functioning simplification' or as a 'disburdening process' apply only to technology-as-a-black-box, that is, an object which has been made unproblematic. However, exactly how an object, an artefact, or a technology has been rendered unproblematic, is the sociologically interesting question. Luhmann makes us think that technology ('trivial machines' in the words of von Foerster 1984) can be taken for granted.

But then, how can air traffic or railway systems, for example, be adequately described in Luhmannian terms? What happens if an airplane crashes, or if a train is late? Does the system (the technology) work 'in a functioning way'? Does it 'relieve the subject'? Is performance possible 'without too much reflection'? These and other questions do not occur to Luhmann, owing to the basic assumption in his theory that technology is not part of society, but belongs to its environment. Recently, however, Luhmann incorporated in his theory a discussion of high-risk technology, which deals precisely with the issue of the boundary between the 'technical' and the 'non-technical' (see Figure 13.1; Luhmann 1990 and 1993b, esp. chapter 5).

Luhmann made an earlier exception for another technological system: printing. Printing has an enormous importance in Luhmann's theory: through printing, Luhmann stressed, knowledge could for the first time be stored in massive quantities, and its spread was accelerated, contributing to the Enlightenment's success.

Electronic data processing was also integrated in the theory, and Luhmann (1989c) drew attention to the scarcely foreseeable consequences that the electronic storage of knowledge would bring. Luhmann acknowledges the possibility that this technology not only perturbs social systems (as other parts of the environment do) but that it directly affects communication: 'the development of these media [distribution media] concerns not only an external increase in communication, it also changes the manner of communication itself' (Luhmann 1995: 162).

Printing and subsequent innovations in information distribution media are of particular interest to Luhmann since they enable communication, beyond face-to-face interaction, between interlocutors who are spatially or temporally separated. It therefore seems appropriate to investigate the *gains in power* which result from *time savings* made possible by technical innovation. Luhmann does once mention this causal chain (Luhmann 1995: 45–6), but he does not further explore its theoretical potential.

Clearly, technologies for Luhmann are *not* social systems, since they are not 'communicative events'. The two exceptions he makes, information distribution media and high-risk technologies, must be considered as *ad hoc* additions to the theory. The first is necessary to explain social change, the

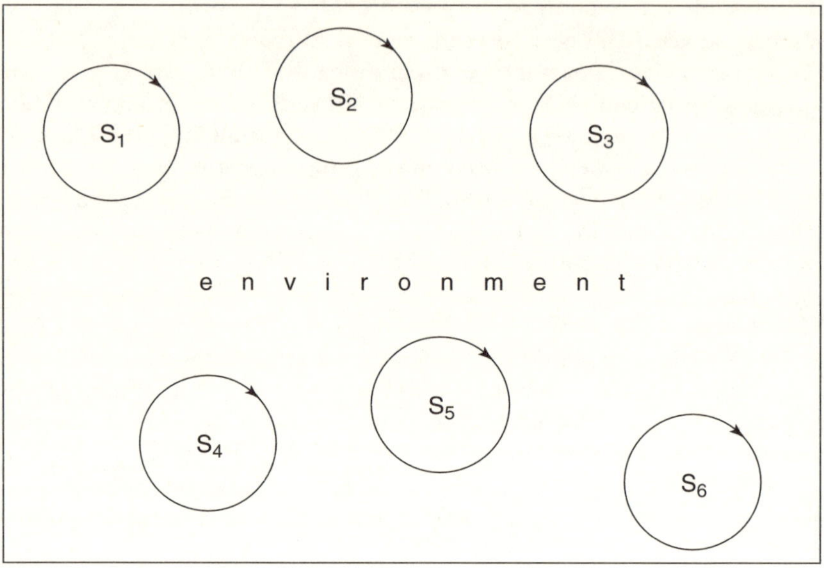

Figure 13.1 Social systems, autopoietically closed

latter is necessary to address ecological problems (for an extensive discussion, see Grundmann 1998).

Large technical systems

> Systems in many respects resemble machines. . . . A system is an imaginary machine invented to connect together in the fancy those different movements and effects which are already in reality performed . . . It often happens that one great connecting principle is . . . found to be sufficient to bind together all the discordant phenomena that occur.
>
> Adam Smith

Scholars studying so-called 'large technical systems' (LTSs) take the opposite approach to Luhmann. They conceptualize complexes of technical artefacts as systems. In so doing, they seem to follow the cybernetic approach, which according to Ashby is the study of 'all possible machines'. Social studies of LTSs, like cybernetics, do not discuss what an object 'is' but what it 'does', what relationships it is embedded in. The classic distinction between the 'social' and the 'technical' is often abandoned, as the following properties of LTSs illustrate: 'they support or sustain the functioning of many other technical systems the organization of which they link up'; 'they are closely coupled technically to complex mandatory organizations and management forms that are imposed by the operational demands of the system itself'; 'they are the object of public anxiety with regard to their safe functioning'; 'they develop partly behind the backs of system constructors and are precarious'; 'they develop in accordance with evolutionary patterns of growth, stagnation, and decay'.[4]

The principal differences between the theory of social systems and the LTS approach are thus apparent: in the former, communications are the only type of system components, whereas in the latter artefacts also qualify; in the former, control is impossible, whereas in the latter it is essential.[5]

We may recall that, in Luhmann's view, technologies are part of the environment of social systems. The impossibility of control among social systems thus does not apply to technologies: social systems integrate technologies according to their needs. The LTS approach, in contrast, conceptualizes organizations *within* technology, social systems within technical networks (Joerges 1996). This makes the steering of LTSs a less easy, straightforward task. Control issues are, again, brought to the fore.

Mayntz and Schneider use the term 'governance' to designate the mixture of cybernetic control theory and self-regulation: 'Governance is understood as the institutionally structured process of self-regulation of social activities. A governance perspective has close affinities with cybernetic theory and related concepts such as control, regulation and guidance' (Schneider 1991: 19).

Actor network theory

> In contrast to language, technology is not a stable system; it can be described in a scientific way only if it is analyzed as a result of a permanent interference between a practical and a technical system.
>
> Jean Baudrillard

'What ties us all together?' asks Latour (199la: 3), and he provides an answer which departs radically from a sociological tradition in which norms are central to the constitution of social order. In the model that he and others develop (Latour 1987, 1991b, 1993, Callon 1987, Law 1987) and which others support (Hacking 1992, Pickering 1993), society is made stable by means of alliances. Social order comes from the strength of networks in which human and non-human elements are mobilized. Contrary to much of sociological literature, this approach maintains that alliances between humans and non-humans are not only possible but crucial for the cohesion of society. The central concept is the concept of *actant*, which designates both human and non-human actors. Non-human actors embody delegations (or translations) of tasks and functions occur from humans to non-humans in so-called socio-technical networks.

These networks are thus 'hybrids' that can non-longer be classified as 'social', 'technical' or 'natural'. Actor-network theory[6] erases the distinction between nature, society and technology, claiming that all cultures have mixed the three at all times. It is a modern illusion that these can be separated (Latour 1991b, 1993). This view was widely contested by critics (see the exchange between Collins and Yearly 1992 and Callon and Latour 1992) who argued that genuine social analysis cannot do away with the categories of meaning, action and strategy which should be reserved for humans.

Based on a number of case studies, Latour and his colleagues developed a model inspired by semiotics,which tries to operationalize the notion of actant. In this model, known as the 'socio-technical graph' (see Latour *et al.* 1992), the evolution of technical or scientific innovations is represented over time. The elements in these 'innovation networks' are treated as interdependent and replaceable. But in constrast with systems analysis, it is not assumed *a priori* that a change in one element will induce changes in all other elements: the extension of the network (the chain of allies) affects only some previously recruited elements; conversely, if one element changes, there may be consequences for the network's propensity to attract new allies or to keep existing ones.

In Latour's model, technical or scientific innovation, or the development of LTSs mobilize three different resources: society, artefacts and nature. The latter two resources are usually not present in other social science concepts of network (see Burt 1982, Coleman 1990, Cook 1990, Laumann and Knoke 1987, Scott 1992). The advantage of this 'unorthodox' approach lies in its sensitivity to processes which are triggered not only by the dynamics of social interactions, but also by the occurrence of natural (or artificial) facts or events and their specific properties. Each resource may behave in surprising ways, which makes the desired outcome uncertain. The main reason for this is that actors are not endowed with fixed power resources: 'It is this very variation of scale that we want to be able to document, whereby a tiny actor becomes stronger than the strong, but without believing in some *a priori* definition of who or what is strong and who or what is weak'.[7] This has at least some resemblance with the approach of social network theory. Here, too, 'one can never simply appeal to such attributes as class membership or class consciousness, political party affiliation, age, gender, social status, religious beliefs . . . and so on, in order to explain why people behave the way they do' (Emirbayer and Goodwin 1994: 1415).

ANT also emphasizes the point which one of the dictionaries offered earlier, the combination of micro- and macro-analysis:

> The scale change from micro to macro and from macro to micro is exactly what we should be able to document . . . The same innovation can lead us from a laboratory to a world and from a world to a laboratory. Respecting such changes of scale, induced by the actors themselves, is . . . important. . . . Given the tools of network analysis that we have at our disposal, trying to endow actors with a

> fixed dimension as well as a fixed form is not only dangerous, but simply unnecessary.
>
> (Latour 1991b: 119)

Latour does not use the standard technique of social network analysis. He does not employ graph theory, nor does he use the basic concepts of density, centrality and the like. However, he does take from the network semantics the view that there is no fixed line between 'inside' and 'outside' and that there is no limit *a priori* as to what may qualify as an element in the network. There is no distinction *a priori* (if at all) between the system and its environment.

The ANT approach, like the LTS approach, acknowledges the need for the control of network elements. Enrolment is always precarious; links and nodes in the network do not take care of themselves but instead need constant maintenance work. Networks are processes or achievements rather than 'given' relations or structures that are given in the order of things. John Law (1995) recently denounced this approach as managerial, obsessed with limitations, mastery and control. Having contributed to the ANT approach, he now gives prominence to the aspect of diversity rather than similarity in this research field, celebrating a 'choreography of otherness'.[8]

Control issues are thus an important aspect of ANT. The development of large technical networks (like any other innovation project) requires the maintenance of the chain of allies and its extension where possible. Only by mobilizing more allies and resources than competitors, will it be successful in outdoing competing artefacts, technologies and networks. However, the term 'actant' seems to be a problematic category which does not fit well into the sociological canon. This is not meant to be an argument for orthodoxy. On the contrary, by stressing the role of strategic choice, motivation and intention, it makes an even better case for the analysis of the processes of mobilization and enrolment.

Boundaries

Dictionaries distinguish between two conceptions of a system: as a whole with its parts, or as an ensemble distinct from its environment. The first conception is frequently used in a quite unspecific way to designate a set of things or phenomena. It is more or less synonymous with the notion of network. The second conception requires specific assumptions about the nature of the system and of its relations with its environment. Thus defined, a system comprises elements that are interlinked in a certain way, which allows for statements on system properties and system-environment relations.

Social systems theory

Luhmann clearly favours the second conception: the theory of autopoietic systems argues that (new) elements of a system are produced by (existing) elements, thus providing an answer to the difficult question of how the unity of a system is constituted. Social systems are conceived of as self-referential, *self-producing* and self observing systems.

In this conception, system boundaries, system components and system operations can easily be characterized. However, many commentators have suggested that SST has lost all contact with real world phenomena. Even Luhmann himself seems aware of such implications. As I have noted above, the incorporation in SST of high-risk technology may be seen as an effort to re-embed social systems in their material environment.

In order to discuss systems' internal relations and systems–environment relations, Luhmann needs a precise definition of the elements and the boundaries of systems. This sets Luhmann apart from the semantics of networks. We can leave the question open as to whether he develops a poor and sterile 'formal science of systems' (Callon 1987). There can be no doubt that preference for the semantics of systems is expressed because of its greater precision in the definition of elements, relations and boundaries. It is a theory with an appeal to the 'absolute'.[9] Networks, by contrast, do not have boundaries. They are characterized by lines and nodes, by density and clustering.

Large technical systems

The use of the system concept within the LTS discourse is not entirely convincing (see the discussion by Joerges in this volume). An important reason for it is that LTSs seem to have fuzzy boundaries. LTSs, Tom Hughes tells us, 'embody the physical, intellectual, and symbolic resources of the society that constructs them' (Hughes 1983: 2). Or, in more artefactual terms:

> 'system' refers to a technical system, such as an electric transmission system. Sometimes reference is, as noted, to a system with interacting components, some of which are not technical. Centrally directed, interacting institutions and technical components comprise such a system. On occasion, however, the concept of system is used much more loosely. 'System' then means interacting components of different kinds, such as the technical and the institutional, as well as different values.
>
> (Hughes 1983: 6)

This broad conception suggests a very complex set of relations among system components. If we distinguish for purely analytical purposes between technical and social elements (i.e. between systems of machinery and organizations) then there are three main types of relations: (1) between several systems of machinery; (2) between organizations; and (3) between organizations and systems of machinery. Of course, there are also reflexive links (4) between organizations and the links between systems of machinery and (5) between

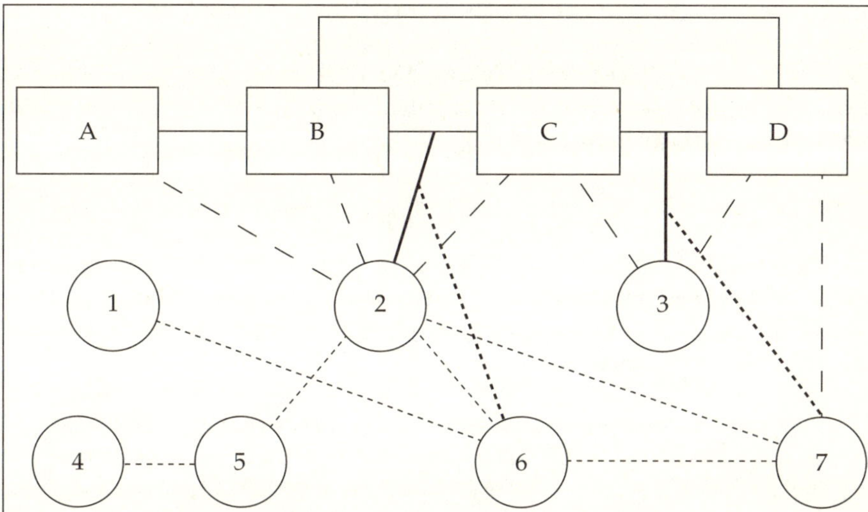

A–D Systems of machinery and artefacts

1–7 Social systems or organizations (e.g. system managers, banks, insurance companies, users, state bureaucracy, operators)

——— Links between systems of machineries (i)

- - - - - - Links between organizations (ii)

— — — Links between organizations and systems of machineries (iii)

——— Links between organizations and the links between systems of machineries (iv)

------------ Links between organizations and the links between organizations and systems of machineries (v)

Figure 13.2 Morphology of large technical systems
Note Hardly any of the components (A, B, etc.; 1, 2, etc.) is 'pure'. The relations between them are complex.

organizations and the links between organizations and systems of machinery, etc. (see Figure 13.2).

According to Hughes, the formation of a technical system is effected through the strategy of a system builder who assembles heterogeneous (and in principle arbitrary) elements into a system unity. This leaves Hughes's approach with the fundamental difficulty of conceptualizing the closure of these systems, for the reconfiguration of technical systems is an ongoing process (see also Grundmann 1994b, Joerges 1996, Mayntz 1988 and 1993). Here one might identify a theoretical problem, for it is not clear what conceptual consequences are to follow. Do LTSs require an 'open systems' approach? Or is the term system applied in a loose way, without clear reference to a specific theoretical framework?

Actor-network theory (ANT)

ANT seems at first glance to be little concerned with the issue of boundaries, insofar as they pose ontological or epistemological questions. Michel Callon's comments on Hughes's conception of systems are revealing of his attitude towards systems theories:

> The system concept presupposes that a distinction can be made between the system itself and its environment . . . How do we define the limits of a system and explain concretely the influence of the environment? To answer such questions precisely, we must develop a formal science of systems, thus possibly depriving the analysis of all its descriptive and explanatory value. Hughes manages to avoid this pitfall by using the systems concept in a pragmatic way.
>
> (Callon 1987: 100f)

The approach taken in ANT is also pragmatic. In fact, ANT simply assumes that anything can be a member of a network. It is exclusively the capacity of the net to mobilize and stabilize new allies. No *a priori* assumptions can be made in this respect. Seen in this way, boundary management becomes the central task of socio-technical networks, for only if the fringes of the network can be prevented from unravelling, will it survive. This pragmatic stance seems reasonable enough. However, the rejection of any a priori assumption need not be a virtue, it can also be a vice, especially insofar as it exempts actor network theoreticians from developing hypotheses.

System and network metaphors, and their implications

In his book *Technology's Storytellers*, John Staudenmaier juxtaposes the cold and hard world of machines and systems with the warm and charming world of the real life. All human aspects like community, norms and values are excluded from the *mechanism*. The systems approach is thus well adapted to the study of technology: 'Because the governing concept of a system is focused on function rather than community, and on hierarchical conformity rather than peer negotiation, it is possible to apply the systems approach to artifacts on many levels of complexity' (Staudenmaier 1985: 71). But it does not answer the needs of social history:

> an exclusive use of the hierarchical systems approach could well dissolve the tension between design and ambience by defining all ambient factors as functional components. Were this to happen, the historical ambience wherein reside personal creativity, aesthetics, politics, and religious faith – in short, all of the passionate, truculent, and unsystematic dimensions of human life and culture – would no longer be visible to the historian.
>
> (Staudenmaier 1985: 81)[10]

It brings up the question of what is at stake when social scientists choose their root metaphors. Wolfe thinks that 'social scientists reach almost automatically for fabric metaphors to describe society [because] threads can unravel . . . As opposed to industrial products, which have hard-metal properties, social properties can disintegrate if not handled properly' (Wolfe 1993: 165).

This seems to capture an important dimension of the problem. Social scientists spontaneously tend to use metaphors of webs and fabrics when concerned about the dissolving of communities or human interactions. For this reason, it seems clear that the LTS approach is nearer to the ANT approach than to the social systems approach, despite their common usage of systems terminology. ANT and LTS both orientate our attention towards shifting boundaries, contested grounds, manipulation of resources and of competitors, be they potential or real. *There is no longer one machine which can be optimized.*

However, the rejection of control issues by Staudenmaier and others is countered by many LTS scholars. They point out that 'we' have to be able to control large technical systems – otherwise we will experience disasters.[11] That control be desirable, however, does not mean that it is always possible, and, when possible, that it will be used only to enhance the *technical* performance of the system: it might as well be used to control people. And success of control is not guaranteed: the old optimistic project of regulating these systems has long gone.[12] 'Every good regulator must be part of the system he wants to regulate' is a slogan of second-order cybernetics and of the self-organizing paradigm (see von Foerster, 1981).

System behaviour obviously depends on the type of coupling involved. In tightly coupled systems, a change in one system component will have direct repercussions on all other system components, especially if these components are related in a complex rather than a linear way (Perrow 1984). Systems divided into subsystems (modular structure) and loosely coupled are more 'error friendly'. They are less vulnerable to shocks (i.e. a change in one system component does not have repercussions on the whole system); but they are also less responsive to attempts at control (von Weizsäcker 1990).

Engineers and system designers are well aware of these facts; but deciding how safe is safe enough, or how much redundancy and loose coupling society can or will afford, is a social process. Politics, partly competing with other social actors and systems, has to bear a large part of the responsibility. Another powerful system is the law, which defines liabilities in system failures. The law can thus stimulate preventive measures which lead to better (safer) system design in the first place. Others actors, such as professional organizations, standard setting institutions and social movements (to name a few), also try to govern these systems according to their interests.

The LTS approach has set itself the formidable task of examining the conditions under which attempts at control can be successful. My analysis has shown that the LTS approach combines central notions from both the systems and the networks semantics. From the former it takes the control issue, from the latter the interrelatedness of *some* elements which are often heterogeneous. It gives up other notions of systems theory central to authors like Luhmann (openness vs closure, system vs environment). It reckons that boundaries are fuzzy and interventions difficult (but not hopeless, as Luhmann would have it).

To come back to the initial question: is the choice between concepts guided by their normative properties? There can be no unambiguous answer to that question. To be sure, the concern over such questions is an ongoing process, as the reaction of such different authors as Buckley, Habermas, Galison and Staudenmeier testify. Their critical attitude to control models and theories are rooted in a longstanding suspicion. Ironically, systems theories in contemporary sociology do not rely on the notion of control. On the contrary, the premise of such theories is that control is not available (whether it is desirable or not). Network approaches on the other hand, despite the fact that they stress the fuzziness of their research objects, are concerned with issues of control. This is not due to the fact that they want to construct precise theories where the objects of research have clear boundaries. It is rather due to the fact that the real life properties of such networks demand proper control in order to prevent disasters: either failure in a competitive environment, or technical failure with disastrous consequences.

This finding should be reason enough to rethink an old idea which has become so dear to many sociologists, philosophers and political scientists alike: that system theory lends itself easily to technocratic social engineering whereas network approaches have an elective affinity to interaction and institutional change.

Notes

1 The German dictionary *Grundbegriffe der Soziologie* mentions Simmel (1908) as one of the founders of the concept of network.

2 The close connection to the Second World War made cybernetics from its outset a contested concept. Not surprisingly, some authors felt uneasy because cybernetics was 'springing out of the last war effort' (Buckley 1967: 36), others going still further and rejecting the whole project for this very reason (Galison 1993).

Bertalanffy is eager to rebuff such preoccupations by emphasizing that there have been precursors of the theory outside the 'war effort', and among them a Swiss (Bertalanffy 1968: 16).

3 Social systems are not entirely closed systems however; they are open with respect to information (cognition). Hence Luhmann's slogan: autopoietic systems are operationally closed, but cognitively open.

4 Quoted from Hughes 1983 and 1987, Joerges 1996, Mayntz and Hughes 1988, La Porte 1991, and Summerton 1994.

5 Issues of safety and security are emphasized by Perrow (1984) for 'freestanding' technical systems like plants and factories, and for LTSs by La Porte and Consolini (1991) and by Joerges (this volume).

6 The label 'actor network theory' seems misplaced since the meaning of the word 'actor' differs from both lay and social science accepted meanings.

7 Latour *et al.* 1992: 45. Also: 'We should never have to presume the stability of either the objects (internalism) or the subjects (externalism); a trajectory is to be defined only by association and substitution of a set of units' (Latour *et al.* 1992: 37).

8 Earlier Amsterdamska (1990) had challenged Latour on similar grounds, arguing that his central metaphors were taken from warfare.

9 As many commentators have pointed out, Luhmann's enterprise can be regarded as a latter-day replica of Hegel's system thinking (see Wagner 1994, 1995).

10 In a similar way, Habermas denounced systems theory as 'social technology', thus alluding to its technocratic properties (see the debate in Habermas and Luhmann 1971).

11 Appalled by Wiener's visions and achievements, the historian of science Peter Galison seeks to distance himself: 'perhaps disorganization, noise, and uncontrollability are not the greatest disasters to befall us. Perhaps our calamities are built more from superorganization, silence, and control' (Galison 1993).

12 Teubner identified a 'regulatory trilemma' where the intervention of a system aiming at regulation of other systems may: (1) cause dysfunctional effects in the latter; (2) remain irrelevant; or (3) cause detrimental effects upon itself (Teubner 1985: 311; see also Buckley 1967: 173–4).

References

Amsterdamska, Olga (1990) 'Surely you are joking, Monsieur Latour!', *Science, Technology, and Human Values* 15: 495–504.

Ashby, Ross (1956) *An Introduction to Cybernetics*, London: Chapman and Hall.

Baudrillard, Jean (1968) *Le Système des Objets*, Paris: Gallimard.

Bertalanffy, Ludwig von (1956) 'General system theory', *General Systems* 1: 1–10.

—— (1968) *General System Theory*, New York: Braziller.

Blackwell Dictionary of Twentieth Century Social Thought (1993) (W. Outhwaite and T. Bottomore (eds)), London: Blackwell.

Buckley, Walter (1967) *Sociology and Modern Systems Theory*, Englewood Cliffs (NJ): Prentice Hall.

Burt, Ronald S. (1982) *Toward a Structural Theory of Action: Network Models of Social Structure, Perception and Action*, New York: Academic Press.

Callon, Michel (1987) 'Society in the making: the study of technology as a tool for sociological analysis', in W.E. Bijker, T.P. Hughes and T.J. Pinch (eds) *The Social Construction of Technological Systems. New Directions in the Sociology and History of Technology*, Cambridge (MA): MIT Press.

Callon, Michel and Latour, Bruno (1992) 'Don't throw the baby out with the bath school: a reply to Collins and Yearley', in A. Pickering (ed.) *Science as Practice and Culture*, Chicago (IL): The University of Chicago Press.

Coleman, James S. (1990) *Foundations of Social Theory*, Cambridge, (MA): Harvard University Press.

Collins, H.M. and Yearly, S. (1992) 'Epistemological chicken', in A. Pickering (ed.) *Science as Practice and Culture*, Chicago (IL): Chicago University Press, 301–26.

Cook, Karen S. (1990) 'Linking actors and structures: an exchange network perspective', in C. Calhoun (ed.) *Structures of Power and Constraint*, Cambridge: Cambridge University Press.

Critical Dictionary of Sociology (1989) (R. Boudon and F. Bourricaud (eds)), Chicago: Chicago University Press.

Emirbayer, M. and Goodwin, J. (1994) 'Network analysis, culture, and the problem of agency', *American Journal of Sociology* 99: 1411–54.

Foerster, Heinz von (1981) *Observing Systems*, Seaside (CA): Intersystem Publications.

—— (1984) 'Principles of self-organization – in a socio-managerial context', in H. Ulrich and G. Probst (eds) *Self-Organization and Management of Social Systems. Insights, Promises, Doubts, and Questions*, Berlin: Springer.

Galison, Peter (1993) 'The ontology of the enemy', *Critical Inquiry* 21 (Autumn): 228–66.

Glanville, Ranulph (1987) 'The question of cybernetics', *Cybernetics and Systems* 18: 99–112.

Granovetter, Mark (1982) 'The strength of weak ties, a network theory revisited', in P. Marsden and N. Lin (eds) *Social Structure and Network Analysis*, London: Sage.

Grundbegriffe Der Soziologie (1992) (B. Schafers (ed.)), Opladen: Leske & Budrich.

Grundmann, Reiner (1994a) 'Über Schienen, Straßen, Sand und Perlen, große technische Systeme in der Theorie sozialer Systeme', in I. Braun and B. Joerges (eds) *Technik ohne Grenzen*, Frankfurt am Main: Suhrkamp.

—— (1994b) 'Car traffic at the cross roads. New technologies for cars, new traffic systems and their interlocking', in J. Summerton (ed.) *Changing Large Technical Systems*, Boulder (CO): Westview Press.

—— (1998) 'Technik als Problem für die Systemtheorie' [Technology as a Problem for Systems Theory], *Swiss Journal of Sociology* 24: 327–46.

Habermas, Jürgen and Luhmann, Niklas (eds) (1971) *Theorie der Gesellschaft oder Sozialtechnologie: Was Leistet Die Systemforschung?* Frankfurt am Main: Suhrkamp.

Hacking, Ian (1992) 'The self-vindication of the laboratory sciences', in A. Pickering (ed.) *Science As Practice and Culture*, Chicago: Chicago University Press.

Heclo, Hugh (1978) 'Issue networks and the executive establishments', in A. King (ed.) *The New American Political System*, Washington (DC): American Enterprise Institute.

Heclo, Hugh and Wildavsky, Aaron (1974) *The Private Government of Public Money*, London: Macmillan.

Homans, George C. (1964) 'Bringing men back in', *American Sociological Review* 29: 809–18.

Hughes, Thomas P. (1983) *Networks of Power. Electrification in Western Society, 1880–1930*, Baltimore (MD): Johns Hopkins University Press.

—— (1987) 'The evolution of large technological systems', in W.E. Bijker, T.P. Hughes and T.J. Pinch (eds) *The Social Construction of Technological Systems. New Directions in The Sociology and History of Technology*, Cambridge (MA): MIT Press.

—— (1991) 'An historical overview', in T. La Porte (ed.) *Social Responses To Large Technical Systems*, Dordrecht: Kluwer.

International Encyclopedia of the Social Sciences (1968) (D. Sills (ed.)), London and New York: Macmillan and Free Press.

Joerges, Bernward (1988) 'Large technical systems: concepts and issues', in R. Mayntz and T. Hughes (eds) *The Development of Large Technical Systems*, Frankfurt am Main: Campus.

—— (1995) 'Prosopopoietische Systeme', in G. Bechmann, J. Halfmann and W. Rammert (eds) *Jährbuch Technik und Gesellschaft (vol. 8)*, Frankfurt am Main: Campus, 31–48.

—— (1996) '"Grosse technische Systeme" oder: was ist groß und was ist klein?', in B. Joerges (ed.) *Technik, Körper der Gesellschaft. Arbeiten zur Techniksoziologie*, Frankfurt am Main: Suhrkamp, 145–71.

Jordan, Grant (1990) 'Sub-governments, policy communities, and networks', *Journal of Theoretical Politics* 2: 319–38

Kenis, Patrick and Schneider, Volker (1991) 'Policy networks and policy analysis: scrutinizing a new analytical toolbox', in B. Marin and R. Mayntz (eds) *Policy Networks. Empirical Evidence and Theoretical Considerations*, Frankfurt and Boulder (CO): Campus and Westview Press.

La Porte, Todd R. (ed.) (1991) *Social Responses to Large Technical Systems: Control or Anticipation*, Dordrecht: Kluwer.

La Porte, Todd R. and Consolini, Paula M (1991) 'Working in practice but not in theory: theoretical challenges of "high reliability organizations"', *Journal of Public Administration Research and Theory* 1: 19–47.

Latour, Bruno (1987) *Science in Action. How to Follow Scientists and Engineers through Society*, Milton Keynes: Open University Press.

—— (1991a) 'The impact of science studies on political philosophy', *Science, Technology, and Human Values* 16, 3: 3–19.

—— (1991b) 'Technology is society made durable', in J. Law (ed.) *A Sociology of Monsters. Technology, Power and the Modern World*, London: Routledge, 103–31.

—— (1993) *We Have Never Been Modern*, Cambridge (MA): Harvard University Press.

——, Mauguin, Philippe and Teil, Geneviève (1992) 'A note on socio-technical graphs', *Social Studies of Science* 22: 33–57.

Laumann, Edward O. and Knoke, David (1987) *The Organizational State*, Madison (WI): The University of Wisconsin Press.

Law, John (1987) 'Technology and heterogeneous engineering: the case of Portuguese expansion', in W.E. Bijker, T.P. Hughes and T.J. Pinch (eds) *The Social Construction of Technological Systems. New Directions in the Sociology and History of Technology*, Cambridge (MA): MIT Press.

—— (1995) 'Traduction/trahison: notes on actor network theory', paper presented at the ERASMUS/EASST workshop on 'Social theory and social studies of science', Bielefeld (9–13 May).

Luhmann, Niklas (1982) 'The world society as a social system', *International Journal of General Systems* 8: 131–8.
—— (1986) 'The autopoiesis of social systems', in F. Geyer and J. van der Zouwen (eds) *Sociocybernetic Paradoxes*, London: Sage, 172–92.
—— (1988) 'The unity of the legal system', in G. Teubner (ed.) *Autopoietic Law*, Berlin: de Gruyter.
—— (1989a) *Ecological Communication*, Cambridge: Polity Press.
—— (1989b) 'Politische Steuerung: Ein Diskussionsbeitrag', *Politische Vierteljahresschrift* 30: 4–9.
—— (1989c) 'Kommunikationswissen und Gesellschaft', in W. Rammert and G. Bechmann (eds) *Jahrbüch Technik und Gesellschaft (vol. 5)*, Frankfurt am Main: Campus.
—— (1990) *Die Wissenschaft der Gesellschaft*, Frankfurt am Main: Suhrkamp.
—— (1992) *Beobachtungen der Moderne*, Opladen: Westdeutscher Verlag.
—— (1993a) *Das Recht der Gesellschaft*, Frankfurt am Main: Suhrkamp.
—— (1993b) *Risk: A Sociological Theory*, New York: Aldine de Gruyter.
—— (1995) *Social Systems*, Stanford (CA): Stanford University Press.
Marin, Bernd and Mayntz, Renate (eds) (1991) *Policy Networks. Empirical Evidence and Theoretical Considerations*, Frankfurt am Main and Boulder (CO): Campus and Westview Press.
Mayntz, Renate (1988) 'Zur Entwicklung technischer Infrastruktur Systeme', in R. Mayntz *et al.* (eds) *Differenzierung Und Verselbstandigung*, Frankfurt Am Main: Campus.
—— (1993) 'Große Technische Systeme und ihre gesellschaftstheoretische Bedeutung', *Kölner Zeitschrift für Soziologie und Sozialpsychologie* 45: 97–108.
—— and Hughes, Thomas P. (eds) (1988) *The Development of Large Technical Systems*, Frankfurt am Main: Campus.
Perrow, Charles (1984) *Normal Accidents*, New York: Basic Books.
Pickering, Andrew (1993) 'The mangle of practice: agency and emergence in the sociology of science', *American Journal of Sociology* 99: 559–89.
Powell, Walter W. (1990) 'Neither market nor hierarchy: Network forms of organization', *Research in Organisational Behaviour* 12: 295–336.
Radcliffe-Brown, Alfred A. (1940) 'On social structure', *Journal of The Royal Anthropological Institute* 70: 1–25.
Scharpf, Fritz W. (1989) 'Politische Steuerung und politische Institutionen', *Politische Vierteljahresschrift* 30: 10–21.
Scott, John (1992) *Social Network Analysis: A Handbook*, Los Angeles: Sage.
Schneider, V. (1991) 'The governance of large technical systems', in T. La Porte (ed.) *Social Responses to Large Technical Systems: Control or Anticipation*, Dordrecht (NL): Kluwer, 19–42.
Simmel, Georg (1971, first edn 1908) 'Group expansion and the development of individuality', in Donald N. Levine (ed.) *Georg Simmel on Individuality and Social Form*, Chicago: Chicago University Press.
Social Science Encyclopedia (1985) (A. Kuper and J. Kuper (eds)), London: Routledge and Kegan Paul.
Staudenmeier, John (1985) *Technology's Storytellers. Reweaving the Human Fabric*, Cambridge (MA): MIT Press.

Summerton, Jane (ed.) (1994) *Changing Large Technical Systems*, Boulder (CO): Westview Press.

Teubner, Gunther (1985) 'After legal instrumentalism? Strategic models in post-regulatory law', in G. Teubner (ed.) *Dilemmas of Law in the Welfare State*, Berlin: De Gruyter.

Wagner, Gerhard (1994) 'Am Ende der systemtheoretischen Soziologie: Niklas Luhmann und die Dialektik', *Zeitschrift Für Soziologie* 23: 275–91.

—— (1995) 'Differenzierung als absoluter Begriff? Zur Revision einer soziologischen Kategorie', *Zeitschrift für Soziologie* 25: 89–105.

Weizsäcker, Ernst Ulrich von (1990) 'Geringere Risiken durch Fehlerfreundliche Systeme', in Mathias Schutz (ed.) *Risiko und Wagnis: Die Herausforderung der industriellen Welt*, Pfullingen: Neske.

Wiener, Norbert (1948) *Cybernetics*, New York: Wiley.

Willke, Helmut (1987) 'Observation, diagnosis, guidance. A systems theoretical view on intervention', in K. Hurrelmann, F.-X. Kaufmann and F. Losel (eds) *Social Intervention: Potential and Constraints*, Berlin: De Gruyter.

Wolfe, Alan (1993) *The Human Difference. Animals, Computers, and the Necessity of Social Science*, Berkeley (CA): University of California Press.

14

HIGH VARIABILITY DISCOURSE IN THE HISTORY AND SOCIOLOGY OF LARGE TECHNICAL SYSTEMS

Bernward Joerges

The high variability discourse of LTS studies

How do social scientists construct large technical systems and networks?* For a long time, academic studies of technology in the social sciences were not unduly concerned with technical ensembles like LTSs: technology research was not part of the canon, and social sciences outside the history of technology did not engage in technology research on a broad scale until the early 1980s. At that time, Renate Mayntz, Thomas P. Hughes, Todd La Porte, Gene Rochlin and others opened up a new research field: 'large technical systems', or LTSs, 'the modern transportation, communication and supply systems, which one might subsume under the heading infrastructural systems, since their primary function consists of enabling a multitude of specific activities to take place' (Mayntz 1988: 233). An international research network was developed and cultivated, a series of conferences took place in Berlin (1986); Cologne (1987); Berkeley (1989); Sydney (1991); Vadstena, Sweden (1993); Autun, France (1995). A number of conference publications appeared (Mayntz and Hughes 1988, La Porte 1991, Summerton 1994b and this volume). Empirical and comparative research got underway in various places; something like a research community, with its characteristic 'discourses', evolved.[1]

How do members of this particular research community (which includes historians, economists and sociologists of technology) talk about large technical systems? In providing answers to this, I will cultivate that peculiar form of empiricism which seems to move only within other writings, without trying to say much about 'the systems out there'. One might argue that what counts in research lies behind or under the verbal carpets. Perhaps, but scientists tend to cover 'one and the same thing' with texts of very different fabrics. The following observations aim at an appreciation of (some of the) patterns of LTS discourse and at demonstrating that behind a label signalling conceptual consistency and a common theoretical focus, LTS research appears to be a highly variable discourse. As in other fields of research, notably in the experimental disciplines,

> not only do different scientist's accounts differ; not only do each scientist's accounts vary between letters, lab notebooks, interviews, conference proceedings, research papers, and so on; but each scientist furnishes radically different versions of events within, say, a single recorded interview transcript or a single session of a taped conference discussion.
>
> (Mulkay and Gilbert 1992: 312)

The following is not meant, though, as another view from above, i.e. a series of meta-statements on LTS research. Rather, I present a side-on view of some discursive practices within a field somebody has aptly called a 'floating seminar'. I will concentrate on the contributions from two volumes on the subject: Mayntz and Hughes (1988) and La Porte (1991), but I will also take into account other relevant texts (Hughes 1987, Weingart 1989, Mayntz 1993, Braun and Joerges 1994b, Summerton 1994b). I shall use these texts to compile characteristic formulations and weave them into 'intertexts', which will in turn be used as a basis for further observations.[2] All quotations are identified at the end of each intertext.

Quasi-definitions and quasi-generalizations

Traditional methodology has it that an iterative research process, which continually doubles back on to itself, should begin with the work of definition, taxonomy and classification, and end with generalized statements regarding empirical relationships. Accordingly, I shall begin with a note on these matters. Later, I will deviate from standard methodological assumptions and examine at greater length certain metaphorical aspects and narrative structures of large technology discourse.

How do LTS authors deal with the problem of basic terminological attributions; how do they approach definitional work? Definitions do more than just determine and protect language. They are powerful filters for distinguishing what should be considered of scientific significance or irrelevance. Often, they do not so much delineate as exclude a particular object, make taboo certain aspects of reality and proscribe the language acceptable in describing them.

Basic terms

Interestingly, there is not much space devoted to clarifying the concept of system, as the first little intertext shows.[3]

> The different components of it form a system: they are all needed and they interact. – Because components of a technological system interact, their characteristics derive from the system. – Social scientists' discussion and efforts towards a more precise analytical conceptualization proceed on the basis of such 'systems' as are characterized by network-like structures, geographic distribution and a considerable concentration of capital. They are thus primarily characterized by the interaction of economic, political and technical-scientific systems. – Although very many empirical systems incorporate some kind of technology, only a subclass of these are organized around a particular form of technology. In cases where one can theoretically eliminate the technology without necessarily implying the absurdity of the remaining organization, the system involved is not a large technical system in the sense applied here.
>
> (Thomas 1988: 179, Hughes 1987: 52, Weingart 1989: 175, 179)

These are the few attempts to explicate the systems term I found in the Mayntz and Hughes and La Porte volumes.[4] Thus, the concept of system as such does not receive much attention, perhaps on the assumption that certain powerful preconceptions associated with it (see below à propos the systems metaphor) do the job.

Somewhat more attention is devoted to circumscribing the distinguishing criteria of *large* (as opposed to smaller) and *technical* (often as opposed to social) systems.

Large systems/small systems

> Thus, the two characteristics of those systems which we shall here define as 'large' technical systems, for lack of a better term, are: 'technical specification' in connection with the interaction of technical artefacts and social organization. Both characteristics indicate that we are concerned with systems where the technology is 'expert intensive' and whose operation is thus highly professionalized. – The system is a large system because of its sheer size in manpower and capital, and because in an advanced state of development, it encompasses most of the territory of a society. – Since the system in this case is extremely large (including one firm which was until recently the largest private business corporation in the world) . . . – And there was a paradox. While the system as a whole contracted,

> technological innovation continued at a rapid pace. – From the technical means of transaction we can distinguish social media of transaction . . . Technical transaction media refer to wires, microwaves or laser.
>
> (Weingart 1989: 180, Thomas 1988: 179, Galambos 1988: 135, Salsbury 1988: 61, Schneider 1991: 22)

In LTS discourses, classifications and basic differentiating criteria tend to remain close to empirical material and are often adopted from practical usage. Certain phrases serve to make it more plausible that LTSs do in fact form a passably circumscribed, definable object, conducive to collective analysis. In the first place, concrete LTSs are often equated with the large corporate organizations and monopolistic utilities which develop and operate them. This creates the problem of losing sight of the incredibly vast fields of activities invested by actors other than the large operating organizations: users, pre-service operations, research and development, related technical structures, etc.

Lacking generally accepted criteria, LTS studies at times simply fall back upon the colloquial assertion that LTSs are unique structures. And yet a sense of uneasiness remains palpable:

> Admittedly, the boundaries are fluid . . . but extreme examples that could be subsumed under the heading of 'large-scale technologies' highlight what is meant here. – [These] are only some of the more spectacular examples of the close collaboration between science, technology, and politics in the implementation of megatechnology. – Efficient regulation has to take into account that LTSs are unique systems.
>
> (Weingart 1991: 8, Weingart 1989: 10, Andersen 1991: 57)

The largest common denominator in the discussion on common characteristics and classification criteria is probably the notion that LTSs might best be described as *multi-actor systems* – whether in the sense of groups assuming certain roles within organizations viewed as enveloping LTSs or in the sense of interorganizational networks.

> [T]he 'ensemble' of what can be termed the 'system of . . . ' . . . consists in the activities of essentially three kinds of actors: First the operator, then the manufacturers, and last the state, in its role as legislator. – First and above all: The development of the system is driven by decisions of a limited number of actors. If a certain amount of momentum developed, this is not a result of forces inherent in an autonomous technology but of purposive action constrained by the sediment of previous decisions about technological alternatives.

> – The motive force of this first wave of spatial integration is an interaction between actors at three distinct levels. At the central state level it is (X). There are regional actors, such as (Y). They often cooperate with their political allies at the local level, in the city halls.
> – Three groups of actors are involved: engineers eager to create, senior executives concerned with safeguarding the rights of the State and lastly, the Parisian bankers.
>
> (Bertho-Lavenir 1988: 155, Thomas 1988: 208, 187, Caron 1988: 72)

Technical systems/social systems

A recurrent issue in LTS discourse has to do with the distinction of and the boundaries between 'technical' and 'social' characteristics of systems (or between technical and social systems).

> Technological systems contain messy, complex, problem-solving components. They are both socially constructed and society shaping. Among the components in technological systems are physical artefacts, such as the turbogenerators, transformers, and transmission lines in electric light and power systems. Technological systems also include organizations, such as manufacturing firms, utility companies, and investment banks, and they incorporate components usually labelled scientific, such as books, articles, and university teaching and research programs. Legislative artifacts, such as regulatory laws, can also be part of technological systems. Because they are socially constructed and adapted in order to function in systems, natural resources, such as coal mines, also qualify as system artifacts.
>
> (Hughes 1987: 51)

This much-quoted passage has proved a very influential, quasi-definitional formulation. Hughes's exhortation not to forget the 'social' components of 'technological' systems is invoked almost ritualistically at the outset of many studies concerned with materialized technology. Elsewhere I have referred to this as the 'Tom Hughes paradox in the social study of technology'. The paradox lies in the fact that sociologists seem to need a reminder that technology consists of dams, turbines and grids, too. Thus, Hughes was and is used by sociologists of technology to justify and legitimize approaches in which materialized technologies play only a minor or nominal role.

Another aspect of the quote which I am interested in here is the persistent tendency, despite Hughes's insistence on seamlessness, to perpetuate the categorical distinction between 'social' and 'technical' in LTS research and beyond. In Luhmannian terms, one might say that technology research in the social sciences operates under the code 'techn(olog)ical /social', not 'techn(olog)ical/non-techn(olog)ical'. Symptomatic in this state of affairs are countless formulations in which technical artefacts are declared not to be social 'as such', as though they could be said to exercise 'non-social' functions and carry out 'non-social' operations.

One dominant formula of LTS discourse is then that there are complex multi-actor 'social' systems and complex 'technical' systems; the latter somehow interact with the former (and may therefore be labelled 'socio-technical'), but remain of a different sort.

So much for the definitional level. It seems that it was not deemed important enough to raise it much above the trivial. But why should we expect our authors to start off with elaborate definitions and then, at the end of a research cycle of extended hypothesis testing, arrive at carefully derived empirical generalizations? In practice, researchers quickly move over, long before the empirical harvest is in the barn, to more or less sweeping generalizations. These generalizations, or 'generalities' as I shall now call them, take over the task of delineation not accomplished by preliminary quasi-definitions.

In scientific usage as in everyday speech, it is often difficult to distinguish once and for all between *useful* generalizations, or generalities, and vapid platitudes. Statements may easily move from one category to the other, and much depends upon the context: what theoretical debates are to be elaborated and addressed, what is to remain excluded from investigations, what is the further end of the argument? Let us look at three such generalities of LTS discourses.

Risky systems/uncertain systems

Generalities based on the 'attractive' concept of risk allow LTS research to latch on not only to high quality theoretical but also to all kinds of public debates.[5]

> The objects of public anxiety about the possible widespread loss of capacity and interrupted service (the more effective it is, the more likely the anxiety) . . . The source of alarm about the consequences to users and outsiders of serious operating failures . . . and subsequent public expressions of fear and demands for assurances of reliable operations. – Moreover, they may have 'catastrophe potential', that is to say that a malfunction may have consequences that are potentially uncontrollable in temporal and spatial terms and therefore with implications for the political and economic spheres as well. – The illusion of control has serious implications for other large-scale technical systems that involve considerable potential risk.
>
> (La Porte 1988: 240f, Weingart 1989: 10, Rochlin 1991: 102)

The risk issue was less predominant at the beginning of LTS research (Mayntz and Hughes 1988) than in the third round in the 'floating seminar' (La Porte 1991).[6] The thematic orientation in that phase, 'social responses' to LTSs, prepared the stage for interpretations of public debates on especially risky systems (La Porte 1988, Oster 1991, Andersen 1991) and technology-related catastrophes (Pinch 1991, Rochlin 1991).

Extending generalized risk assumptions to cover all large systems is a certain way of gaining public relevance. When belief in progress wanes and confidence in science and in the controllability of technology falters, then anything that is opaque, complex and large seems necessarily to be perceived as dangerous. The occasional references to environmental problems in connection with ruptures in systems development seem only a side-line here. Perhaps one may even see the 'risk refrain' as a rhetoric actually enabling LTS research to avoid in-depth ecological discussions (see below).

Self-organized systems/self-regulated systems

The risk discourse is clearly a sciences-of-the-artificial discourse. Its roots are in engineering and cybernetics, and its metaphors are technical. But there is another systems theoretical field where systems are, right from the beginning, not conceived as artefacts but as evolving and emerging entities. This brings us to a third generality, which has to with the idea that LTSs may be self-organizing systems. The notion of self-organization seems here to revolve around two controversies. Are LTSs inherently self-guiding, or are they subject to outside control? Are they homeostatic and harmonious, or crisis-ridden, in danger of collapse, chaotic? Note that LTSs are conceived as self-controlled, self-generating, self-enhancing or self-destructive formations from both viewpoints. On the other hand, talk of 'own dynamic', self-regulation and self-organization seems to hold the promise of ambitious theoretical interpretations, treating LTSs as special cases of differentiated self-organizing social systems.

> The system develops a direction and goals and as it grows, it acquires momentum. – The organization also develops a culture which fosters this growth. – There are directions of technical development that are cumulative and self-generating. – As systems mature, they acquire style and momentum. – The high-momentum systems of the inter-war years give the appearance of autonomous technology. Because an inner dynamic seems to drive their course of development, they please managers who wish to reduce uncertainty and engineers who need to plan and design increased system capacity. – Such systems appear to be closed ones, not subject to influence from external facts or from the environment. – Large technical systems have an inherent

> tendency to expand. The question is under what conditions they succeed and under what conditions they fail.
>
> (Salsbury 1991: 86, 86, Hughes 1987: 56, 79, 79, MacKenzie 1991: 165, Weingart 1989: 187)

Such quotations indicate that many authors see LTSs as phenomena whose development, growth and problems can, for the most part, be explained as self-organizing processes. This does not follow directly from the under-specified sense in which the system concept is generally applied. On the other hand, it corresponds well to the equally persistent theme of LTS uncontrollability and gives welcome support to this idea.

The toss-up remains then between the concept of system prevalent in the engineering sciences on the one hand (implying steerability) and the sociological concept of system on the other hand (implying self-organization and supporting theories of uncontrollability). Although the talk of inner-system dynamics, autonomy, self-propelling 'momentum' and poor external influence are part of a standard repertoire, one cannot say that LTS research is dominated by a generalized discourse of self-dynamics and self-control, aimed at an ambitious theorizing of technical systems. Renate Mayntz's proposals in this respect (1988, 1993) have been hesitantly accepted. For terminological and disciplinary reasons, her conceptual challenges did not induce the LTS community to develop a generally accepted systems model in line with a theory of societal differentiation. Although the self-professed target was and is (in general) to come up with generalization across systems, time and space in the end, narrative approaches and case studies have predominated in practice, inviting generalizations on the one hand but excluding any serious move from the general to the specific, on the other.

Metaphors of LTS discourses

In the discourses on large technology which are drawn on here, not only is the system concept itself used metaphorically, but a series of metaphoric leitmotifs is apparent throughout: the hero metaphor as well as the evolutionary metaphor with its variations: the harmony/conflict tension, the technology/economy complex, and the technology-state theme. The imagery used in talk about large technical systems potentially serves as a theoretical resource, in that it creates unlikely analogies between LTSs and something seemingly utterly different, and thus can produce surprising insights. Donald McCloskey (1985) put it beautifully for the field of economics: every model, no matter how abstract, has its origin in metaphoric images. Conversely: metaphors are theories *in nuce*, pretheoretical models. In this sense they are the opposite of (hi)stories, even though they always tend to be accompanied by narrative elements, with their implicit normative pointers – the moral of the story.

The systems metaphor

Since the use of the term system in LTS discourse is virtually never supported by definitional work, the word is either taken for granted in its self-evident connotations (à la Webster: LTSs are called systems because they obviously represent 'regularly interacting groups of items forming unified wholes') – or else, it is used metaphorically. I suggest that the near refusal to elaborate systems concepts implies that the notion of technology as systemic by nature is itself indeed metaphoric. Staudenmaier (1985) has given us a memorable analysis of the varied uses historians of technology have made of the word 'system'. To what extent do our authors rely on a pretheoretical systems metaphor and what could be the consequences?

In most of the statements in which LTS texts mention systems, this occurs in an everyday fashion: every listener or reader knows that one is speaking about something connected, something integrated, without any further specification. But especially where authors stress the expression 'system' in a way meant to signal a special viewpoint, superior to other perspectives, they are using the expression metaphorically in the sense of 'unexpectedly is like . . . ' or 'surprisingly functions analogously to . . . '. Hughes, for instance, used the concept of system in order to point out and celebrate hitherto unacknowledged connections between a multiplicity of technical and other cultural artefacts, and called men capable of establishing such connections on a grand scale 'system builders'. In a retrospection on LTS research, he noted:

> Recently . . . I discovered – too late to correct the misunderstanding – that most of my historian colleagues assumed that the system builders of whom I spoke were and are little more than business entrepreneurs and that their system-building differed and differs little from the founding and development of business firms. It also dawned on me . . . that my colleagues use mechanical metaphors to explain relationships while I use ones borrowed from electrical engineering. My metaphors tend to circuits, fields, and systems; theirs to mechanical trains of cause and effect.
>
> (Hughes 1991: 188)

Hughes's – if I may say so – 'electroform' version of the systems metaphor has left its imprint on much in early LTS research and points to the more general condition that LTS discourse on the whole takes its metaphors from the tradition of systems and control theories in the engineering sciences. LTSs resemble things cyberneticists talk about.

But the observations with regard to the 'self-dynamic' character of large-scale technical systems also show that social science versions – Parsonian, Luhmannian or Mayntzian – are at work as well. LTSs are also social systems of the kind that concern theoreticians of sociological systems. Sometimes an uneasy amalgamation between cybernetic control theory and the theory of self-regulation takes place: 'Governance is understood as the institutionally structured process of self-regulation of social activities. A governance perspective has close affinities to cybernetic theory and related concepts such as control, regulation and guidance' (Schneider 1991: 19).

Upon close examination, it appears that the use of the self-organizing-system metaphor remains, on the whole, without much consequence for further theoretical or policy oriented interpretation of LTSs, and nowhere is the hermetic quality of a (Luhmannian) theory of social systems adopted. Similarly, where one hears slight intonations of a certain systems mystique, it is likely to be a 'first order construct', one that is taken from the talk of system members: '(T)he concept of the network acquired a powerful mystique among system managers and workers . . . One system, one policy, universal service' (Galambos 1988: 141–3); or else: '(A)ll were guided by a progressionist and nationalist ideology and sought above all to create an operational instrument capable of increasing their social influence out of all proportion . . . Seen in that light the system took on symbolical meaning' (Caron 1988: 73).

Heroes and giants

There is something heroic about LTSs. Again, this theme has much to do with Thomas Hughes and his concept of the 'system builder', focusing on a superior power to get things done:

> Because they are invented and developed by system builders and their associates, the components of technological systems are socially constructed artifacts. Persons who build electric light and power systems invent and develop not only generators and transmission lines but also such organizational forms as electrical manufacturing and utility holding companies. Some broadly experienced and gifted systems builders can invent hardware as well as organizations, but usually different persons take these responsibilities as a system evolves. One of the primary characteristics of a system builder is the ability to construct or to force unity from diversity, centralization in the face of pluralism, and coherence from chaos. This construction often involves the destruction of alternative systems.
>
> (Hughes 1987: 52)

The picture of large systems promoted by heroic system builders, historic leaders and grand strategists (a faint echo of Clausewitz) and engaged in titanic struggles with rivalling systems sometimes evokes martial images: couched in a code of domination and submission it comes across as a peculiarly masculine metaphor.

> Outstanding examples of independent inventors and their radical inventions that sowed the seeds of large systems that were presided over by new organizations are . . . – Further German penetration was . . . held up . . . because the French manufacturers were hostile to the practice . . . They feared indeed an invasion of their market. – The monolith is challenged. – The rise and fall of the German system suggests some interesting conclusions. – X himself issued the first telegraphic order to control trains . . . The engineer would not obey the order, and X himself drove the locomotive to the next station. – This sets the stage for a battle between the Y and Z interests for control of the nation's . . . system.
>
> (Hughes 1987: 58, Bertho-Lavenir 1988: 166, Galambos 1988: 146, Heinze and Kill 1988: 131, Salsbury 1988: 44, 45)

Galambos (1991) noted that prominent individuals were strongly represented in LTS case studies, adding that systems are often reduced to their powerful spokespersons for pragmatic reasons (one need only write Mrs/Mr Smith, instead of always giving involved descriptions such as 'the director of the government program for the control of . . . '). To this extent, one should not overestimate the significance of this characteristic. Personal attribution nevertheless carries another significance: there is a marked tendency to link positive effects, successes and achievements to individuals; failures and negative effects, on the other hand, are easily relegated to the diffuse level of 'the system'. Failure of the system is practically never associated with a name in these studies.

But the heroic image does not only function as a means of celebrating the exceptional role of system builders. In a broader sense, it also stands for the way in which LTSs (or more precisely: the large organizations running them) can be envisaged as powerful 'super-persons'. And sometimes heroic men stand for machines, or heroic machines for men.

Evolution

The evolutionary metaphor has always been popular with historians and sociologists of technology. An easy biologistic rhetoric would have it that LTSs are conceived and born, that they grow, survive and die.

> The American railroad network . . . started its life unaware that it would institute sharp breaks with past business traditions. – The birth of the Bell System. – However, its painful birth made the characteristics and limits of the telephone system clear. – The early technological development in this sector abroad gave Germany the advantage of being able to adopt complete, proven and functioning systems which had already passed their teething troubles. – The

> American railroad network as a vital and progressive large scale technical system reached its apogee in the period between 1900 and 1914. – It [the French railway system] was sure to survive and revenues would suffice not only to meet operating costs but also to guarantee ample return on the capital committed in the vast undertaking. – Because a system usually has embodied in it characteristics suiting it for survival in a particular time and place, manifold difficulties often arise in transfer at another time or to a different environment. – This chapter has dealt with the patterns of growing or evolving systems. Countless other technological systems in history have arrived at a stage of stasis and then entered a period of decline.
>
> (Salsbury 1988: 65, Bertho-Lavenir 1988: 173, Galambos 1988: 136, Heinze and Kill 1988: 116, Salsbury 1988: 61, Caron 1988: 81, Hughes 1987: 87, 80)

One might say that this is common parlance and in this sense to call it metaphorical is giving these texts undue poetic significance. But the metaphor of evolution carries deeper meanings, beyond such almost idiomatic usage. Aside from the affinity to the complexity discourse referred to above, LTSs are often portrayed as generally adaptive, as contingent upon mechanisms of mutation and selection, and as subject to processes of maturation.

> Fixed on his aim, he even opposed 'small steps' that were improvements of the status quo; he did not understand that complex systems have to prefer evolution to revolution because big steps lead to extreme rates of change in other subsystems and thereby endanger the whole system. – What happens is adaptation and not radical change. – The technological mutation taking place between the 1870s and 1900 must therefore be understood first of all in terms of the harmonization of the different branches. – Suppliers are consequently put through a ruthless selection process. – The overwhelming response of the social environment shows that the specific historical situation evokes system innovations and makes the superior system variant the superinnovation of bridging space. – . . . whereas the technical system proper attained a degree of maturity that ensured its efficient working in economic terms.
>
> (Heinze and Kill 1988: 116, Bertho-Lavenir 1988: 164, Caron 1988: 96, 86, Heinze and Kill 1988: 107, Caron 1988: 70)

One may feel that the evolutionary metaphor does not agree well with ideas of planning and control implicit in the 'system-builder' concept and in other heroic metaphors. But, at least in the Darwinian version which apparently underlies most uses of evolutionary metaphors, one finds that a series of impersonal mechanisms are postulated ('competitive struggle', 'adaptation', 'mutation/ diversification') which strategic action might conceivably latch on to. Indeed, the history of social Darwinism shows that the metaphorics of evolution are flexible to an extraordinary degree; they can serve as a vehicle for very different views of society and politics. In a recent essay, Amartya Sen (1992) argued that evolutionary thinking can provide the basis both for pessimistic or fatalistic positions and for humanistic reforms. But the tension remains and, as will be demonstrated below, generates different (or contradictory) stories of 'the development of large technical systems'.

Evolutionary theory has developed, too, since Darwin. The old dispute over whether natural history advances 'gradually', in measured steps, or whether it moves 'catastrophically', in erratic forward (and occasionally backward) leaps, is still going on in theories of 'punctuated equilibria' and the like. The LTS field seems to support the gradualists in this respect. Not so with regard to another seasoned controversy in the general theory of biological and other systems: do systems seek to attain a state of equilibrium or do they, on the contrary, usually operate far from it? In social systems theory, this controversy was tantamount to a debate over the explanation of social change. In the following section I will argue that many LTS researchers adopt 'catastrophic' stances in this regard.

Whatever the nuances, evolutionary metaphors accomplish the important task of *naturalizing* LTSs: to posit them as part of a natural social order, amenable to objective analysis, not as cultural and discursive artefacts open to constant reinterpretation.[7] In doing so, the young specialty of LTS studies (and social science technology research more generally) can follow the lead of mainstream economics of technology.

Harmony and equilibrium/crisis and conflict

Here, LTS research finds itself in considerable conflict. Do systems evolve towards harmony and equilibrium?

> The technological mutation . . . must therefore be understood first of all in terms of the harmonization of the different branches. – This required a harmonization of the different components of the system. – The days of trial and error end . . . when the institutional system achieves an equilibrium, albeit fragile, that can no longer be challenged. – . . . system leaders try to preserve harmony and peace and overlook ways to make money. – [T]he system, which had comparatively good labor relations at the time and which was making satisfactory profits, decided not to let technological change upset a care-fully worked out harmony.
>
> (Caron 1988: 96, 102, 70, Salsbury 1988: 63, 63)

Or are they going through a 'perpetually refuelled crisis' (Bertho-Lavenir (1988: 158), plagued by conflicts and the permanent threat of immanent collapse?

> The system . . . was in a state of crisis by the year x and collapsed altogether after the . . . revolution. – When systems operations collapsed . . . the companies had no means of adjusting the numbers of their staff to cope with the new situation. What was more, the previous organization into skilled and functional technical departments was rusty. – The history of the system, in fact, is dominated by a 'crisis' situation. – Only with the collapse of the giant agency . . . did its managers, as well as government and union leaders, begin to question the old managerial system.
>
> (Caron 1988: 76, 100, Bertho-Lavenir 1988: 158, Salsbury 1988: 66)

A toss-up, again. The way the systems metaphor is used and reified tends to place issues in a homeostatic frame. And yet, observation of concrete systems clearly suggests that most move from one precarious state to another most of the time. LTSs are likened to biosystems then, they evolve.

At another level, the meanings of LTSs and the explanations offered for their emergence revolve around a whole series of oppositions which I call metaphorical because in all these cases core concepts from one competing subdiscipline in the interdisciplinary field of LTS research, say economics, are taken up by another subdiscipline, say political analysis, in a metaphorical way. Central concepts from one field are borrowed and exploited as highly persuasive focusing devices for observations and arguments that have to a large degree been generated in quite different fields. I will shortly point to two such crossovers which are particularly rich in their implications for the further elaboration of mechanisms of system generation as well as LTS policies: the tendency to view LTSs as primarily (essentially) economic systems as opposed to claiming some essentially extra-economic substance for them; and the manifold views of the relationship between LTSs and (nation) states construed in terms of their similarities and dissimilarities.

Economy/technology

'All LTSs are wholly economic!,' say some; 'they are technical, as the name says!,' reply others. Taking this particular instance of the technical/social divide (which has parallels), how is the interrelationship between economic and technical efficiency construed? What are the metaphorical resources tapped in playing the economic versus the technological card? Note too, in the following passages, the ongoing subtext about the equilibrium/conflict between them.

> Even though technological constraints might prevail in the end, it is hard to believe that economic forces . . . do not play a part in the emergence of the system and do not eventually shape its structures. – This combination of 'imperatives' leads to a fundamental and abiding tension between technical safety and reliability on the one hand and economic efficiency on the other . . . – This situation is underlined . . . by the premature shutdown of system components whose construction is a scientific-technological success, but a disaster in economic terms. – System component designers must make trade-offs among conflicting demands of producing a component that is easy and inexpensive to construct, inspect, maintain, and operate. To be a commercial success, the component must also perform a wide variety of missions in terms of . . . loads, distances, and operating environment with a high degree of scheduling reliability. Both unnecessary cost resulting from overdesign and inadequate reliability from underdesign can lead to . . . disaster.
>
> (Lévy-Leboyer 1988: 261, La Porte 1988: 225ff., Weingart 1991: 9f., Oster 1991: 127)[8]

At the first LTS conference in Cologne, there was too much talk of technical efficiency, 'load management' and the like for the taste of Charles Perrow. Turning to Tom Hughes, he called out, 'It's economics, economics are the mountains!' But if economics are the mountains, if the scale and scope of a technical network are determined by its economic trimmings, then, one might say, technology is the invisible part of the iceberg. However, determining conceptually (let alone measuring) what proportion of the dynamics of non-technical (for example economic) system parts is due to dynamics of a technical origin, proves difficult. At this level, LTS discourse sounds suspiciously like the jargon of system managers: caught in the self-made trap of categorically distinguishing between 'social' (in this case economic) and 'technical' systems, one falls back, in most attempts at empirical specification, on parasitizing the discourse of actors in the system.

State-technology/technology-state

Another semantic field is occupied by a set of metaphors having to do with the affinities between states and LTSs. In a certain sense, some authors seem to be telling us, states are like LTSs and LTSs are like states, they have a lot to do with one another and cannot exist without each other. But again, high variability is at work. The next intertext weaves together certain references alluding to political theories and convictions 'acting in the background' (and maybe sometimes behind the backs of their authors . . .).

> From the very start, system development was contained within the

> bounds of rigid administrative regulations. – The companies lost the control of rates. They were prohibited from signing private agreements . . . They were subjected to financial control and they had to obey the rules of administrative accounting. The price paid for the monopoly, the extension of the network and the investment security was administrative tutelage . . . – The . . . crisis merely served to highlight the operational difficulties of the system. State tutelage had become too burdensome . . . The State levied heavy charges on the companies without compensation . . . – In the process of policy formation and the subsequent allocation of regulative burdens, business associations and semi-public institutions are often used as mechanisms for interest intermediation.
>
> (Caron 1988: 78, 81, 84, 99, Schneider 1991: 34)

LTSs are what they are by virtue of state regulations, and the state enacts regulations whereby it profits from the systems. LTSs and the state are closely interdependent, and the relationship is *parasitic*: the state regulates and reaps the benefits, free-riding as it were. At times, however, the state also impedes LTS development. It is power-crazy and avaricious, and whenever it intercedes through regulations, it does not act for the benefit of the LTSs. LTSs thus appear as typical state-technologies.

Yet there is also another, friendlier side to the technology-state:[9]

> Only the State could build the system because it alone sought no return on its capital. – The government systems engineers do everything they can to promote the emergence of a national industry. – With considerable encouragement from . . . industry, the Federal government reluctantly accepts responsibility for licensing operators, inspecting equipment and supervising the use of local installations and operating safety. – By a series of legislative acts, the French state has assumed . . . complete control over the development of the system, and over the procedures preparatory to building new major components. – The existing state system monopoly offers the central government a focal role in the introduction of the new technological system from the start, and the existing . . . network makes the plan of a nation-wide extension of the new service feasible.
>
> (Caron 1988: 74, Bertho-Lavenir 1988: 156, La Porte 1988: 216, Lévy-Leboyer 1988: 259f, Mayntz and Schneider 1988: 264)

LTSs are what they are by virtue of state cooperation and promotion; each is dependent on the other; the dependency is *symbiotic*.

Inextricably interwoven within this discourse, which should be termed metaphoric on account of its likening two apparently very dissimilar phenomena, is a subtext having to do with processes of nationalization/

privatization. This theme, which continues into the narratives of responsibility taken up below, has certainly gained prominence in the course of the LTS conferences (see in particular Salsbury 1994: 142–59, Rochlin 1994: 239–55, Radkau 1994: 64–88, Kubicek 1994: 112–24, Ekardt 1994: 171–88).

To sum up these observations concerning LTS metaphors: there is an amazing variety of interpretations, constructions and stories *in nuce* flourishing around the notion of Large Technical Systems. A few recurring root metaphors stand out more distinctly; others may easily be traced as well. Particularly the examples from the semantic field of technology/economics/state show that here, too, one can adopt Thomas Hughes's memorable phrase and speak of 'seamless webs'. This image was conceived to characterize the systems 'out there', but it is also a good description of the way large technical systems are treated in the history and sociology of technology.

Narratives of LTS discourses

At first sight, most LTS studies may read like relatively straightforward case studies or historical reconstructions of concrete systems, simply telling how it all happened. But as with the level of metaphoric imagery, several distinct narrative structures are at work. Two patterns, in particular, recur throughout the multiplicity of accounts of concrete technical networks: the division into stages, phases, eras and ages of development, i.e. stories about the orderly nature of LTS growth processes; and the attribution of responsibility, i.e. stories of who or what steers LTSs.

Stages and ages

Almost all LTS studies, especially those written by historians, are arranged in a series of stages more or less explicitly accounted for. Stage-devices (still close to model-building based on certain metaphors) and their attendant periodizations (closer to narrative and storytelling than to models) belong to the oldest tools of the historical sciences. They help in achieving the transitions from straightforward, associative (or syntagmatic) narratives – 'and then . . . and then . . . and then' – to substitutive (or paradigmatic) – 'or . . . or . . . or' –, i.e. to generalizations with reference to many other, similar cases.

Thomas Hughes (1987) has set the tone here, again, in proposing a basic scheme of LTS-stages as a frame for ordering and making sense of events across many systems and system types. The three developmental stages, which he had established studying the American, British and German electricity generation and distribution systems, comprise: 'invention and development'; 'innovation and competition'; and 'consolidation and rationalization', a script for a clearly upward and outward reaching movement resulting, if successfully played out, in mature LTSs of high momentum.

> The phases can be further ordered according to the kind of system builder who is most active as a maker of critical decisions. During invention and development inventor-entrepreneurs solve critical problems; during innovation, competition and growth, manager-entrepreneurs make crucial decisions; and during consolidation and rationalization, financier-entrepreneurs and consulting engineers, especially those with political influence, often solve the critical problems associated with growth and momentum. – This dynamic mechanism of success can be represented as a four-step acceleration process: (1) the existing system reaches its capacity constraints; (2) a new technology is at hand; (3) improvements of the old technology and isolated usage of the new technology increase demand for transportation; (4) this additional demand allows the full engagement of the new technology. – An analysis of other large technical systems indicates that this pattern of development is in no way unusual . . . One might therefore conclude that this is a general pattern in the development of successful technical systems serving a specific function. – Looking at the historical development of the network, one can identify four stages of development . . . : invention and isolated introduction (localized linkage), demand-oriented construction (integration) – fulfilling only the needs of existing business centers, supply-oriented extension (intensification) . . . , maintenance-oriented 'cut-back' (selection) . . . Whether system development will be successful is decided in the early stages of the process.
>
> (Hughes 1987: 56f, Heinze and Kill 1988: 129, 105, 105)

Take for instance the development of the French railroad, recounted by François Caron; or the story of United States' telecommunication (from Bell to AT&T), as told by Louis Galambos; or that of the German road network, as presented by Heinze and Kill, or French electrical power grids as described by Lévy-Leboyer – the ease with which these and other narratives of technological history fit Hughes's scheme indicates that this model has a basic narrative structure in itself. It shows the basic plot according to which the development of many regional system networks has so often been described and actually planned. But not always: Mayntz and Schneider (1988) did not see evidence that there is a sequence of such stages operating in the development of government controlled communications systems (like Teletext). Objections were also made on the grounds that various national stories are actually rather different (Lévy-Leboyer 1988; as opposed to Bertho-Lavenir 1988), or, as one would expect from the above-mentioned 'electroform' origins of Hughes's principle of stages, that the development paths of different technologies are not all that comparable.

Stages of development are always timed, social studies of technology always create periodizations. As noted in the context of evolution metaphors, a prevalent idea in the field of LTS research is that of an orderly, directed (if not always planned and sometimes crisis-prone) trajectory from small (young) and local to large (old) and national systems. But this form of periodization proved not entirely defensible, and its initiator Thomas Hughes gradually developed it further. Today one can distinguish three forms of periodization in his work.

The first, developed in *Networks of Power*, reconstructing (energy-) time from 1870 to 1930, had three stages. Maybe we have in many of the stories told in Mayntz and Hughes (1988) for the last time interpretations which hark back to the era of enthusiastic postwar reconstruction, of LTS development as the story of an inventive beginning, a tempestuous development and a stable state of maturity. But already that same year, Renate Mayntz published her 'On the Development of Technical Infrastructural Systems', and concluded that at least in continental European nations, because hierarchical social orders (especially centralized government) and centrally-controlled LTSs are 'congruent', the decline of the former might contribute to that of the latter (Mayntz 1988: 275ff.).

One year later, Thomas Hughes presented his second big study on the development of technical systems (Hughes 1989) which leads up into the 1970s. The stage model and its implicit success story were abandoned here in favour of a story of the rise and fall of large technical systems, in tune with the American self-interpretations of decline so characteristic of the 1980s. The structural-analytical concept of the 'momentum' of large technical systems following the electrical pattern yields to a parable of cyclical generation and decay. LTSs, Hughes stresses, are now culturally threatened, are obsolete and doomed to extinction by the counterculture which emerged in the 1970s, with its vision of a decentralized, soft technology.

As a consequence, Hughes has tended since to concentrate on systems like 'management systems' and the technically hybrid telematic systems of the more recent past. At the same time, he seems now to interpret systems resembling grid-based energy systems as 'modern', and the non-gridbased, hybrid project management, and information systems of the 1970s and 1980s as 'postmodern'.[10] The secular metaphor has also been adjusted: the implied epochs are now distinguished into a modern era, beginning approximately with the last turn-of-the-century, and a postmodern era, drawing, it seems to me , on a pervasive and persuasive fin-de-siècle rhetoric. Be it said that this division of LTSs into modern and postmodern exemplars loses some of its persuasive force because the old systems rhetoric survives in part: reflecting upon systems in terms of complex causal interrelationships has not been abandoned, although it doesn't quite fit into postmodernist rhetoric. (Or else it is another case of high variability talk by one and the same scholar.)

Further interpretation would have to work out in more detail what plots underlie the transitions from one act of the drama to the next: which mechanisms are seen as transforming the systems along their path of development and possible decline?[11] In order to pursue this at least a short way, another master narrative will be traced, which has to do with the question of who actually does – or does not – control technical development. The rise or fall of LTSs in these stories is analysed in terms of achievement or failure for somebody is always held accountable.

Narratives of responsibility: emergence and openness versus closure and design

Narratives allow us to learn from history by providing a moral to the story – something that models do not accomplish, because models have to do with possible, counterfactual, not actual developments. Who, then, are the heroes, and who the villains in LTSs? And do systems evolve by themselves behind the backs of actors, or are they the result of strategic communicative acts? As it turns out, narrative modes closely akin to the story of phases and stages are clearly retained in most studies. They deal with system inventors, system constructors, system designers and system managers intentionally promoting and shaping their systems.

> Networked LTSs are tightly coupled technically, with complex 'imperative' organization and management prompted by operating requirements designed into the system, i.e., unless operations are conducted in x, y ways, there are no benefits, maybe great harm can be imagined. (This is a kind of soft technical determinism: either do it my way or it won't work and do good things for you.) – Such systems reify the models of the designers, imposing a modality of control as if an exhaustive, predictive knowledge base were in place. – Durable physical artifacts project into the future the socially constructed characteristics acquired in the past when they were designed.
>
> (La Porte 1988: 240–1, Rochlin 1991: 118, Hughes 1987: 77)[12]

Peter Weingart indicates how indecisively this entire discourse vacillates between a systems and an actors perspective: almost as a rule, case histories begin with the heroic system builders, who at some point then disappear from the stage, leaving the system to anonymous forces from within or without: 'It is the strategy of the "system builders" to gain control over the "intractable forces" in the system environment, i.e. the factors which generate uncertainty for the system, and to structure the environment to meet the demands of technology and the social systems organized around it.' Weingart continues by saying that in order to secure their assets, 'systems' must eliminate uncertainties; that this in turn means expanding, insofar as a system succeeds in overcoming the opposing forces in its environment or, alternatively, restructuring the environment according to its own imperatives:

> In principle . . . the system is inherently oriented towards bringing its entire relevant environment under its control. It is, however, not only impossible to achieve this goal in principle; moreover, since those parts of the environment which have been brought under control become components of the system, its expansion increases its degree of inner complexity and increases the problem of its internal regulation.
>
> (Weingart 1989: 181f, my translation)

This by now familiar piece of LTS discourse ends with formulations that posit large technical systems as self-regulating entities on their way to inevitable control crises due to the ongoing incorporation of complexity from their environments. Listen to Thomas Hughes again:

> Over time, technological systems manage increasingly to incorporate environment into the system, thereby eliminating sources of uncertainty, such as a once free market. Perhaps the ideal situation for system control is a closed system that does not feel the environment.
>
> (Hughes 1987: 53)

The metaphors behind this (end of) story come mostly from biological and sociological theories of self-organizing, autopoietic systems which tend to exclude notions of external control and steering. Yet the beginning of the story is couched in action theoretical terms: the notion that there are agents designing and controlling systems.

In the course of the debate at one of the LTS conferences, Renate Mayntz insisted:

> [One of the issues] of the research challenge that I see . . . is the design issue, i.e. the question how the anticipation of consequences along the different performance dimensions . . . determine(s) system design. What do system designers really anticipate? Which of the various consequences do they anticipate? And if they anticipate them, how does that enter into the design process?
>
> (Mayntz 1991: 183)

LTS studies provide scarcely any evidence on this point. On the other hand, one suspects that the blithe talk of systems design, of intention and learning, of control through technology and so on provides little in the way of conceptual accentuation: it is almost always found in connection with systems components (airplanes, trains, battleships, supercomputers) and hardly ever with regard to any of the extended networks linking them with so many other things. Remember also the basically sceptical attitude perceptible throughout these studies when it comes to the topic of systems control and controllability.

This scarcely allows for a foregrounding of stories of design and closure, which has a lot to do with responsibilities for outcomes.

So what we have here is perhaps another *façon de parler* in social studies of technology, adopting, as so often happens, elements of the rhetoric of informants 'in the field' and system members who, especially in early stages, have good reasons to present their schemes as well-designed and under control. Inasfar as a more or less casual actors' rhetoric is retained throughout the telling of a system's story, the fundamental openness and uncontrollability of LTS development which is also asserted in most LTS studies is commonly framed as a matter of unintended effects and consequences.

All in all, the notion is pervasive that LTSs are – over and above those components which can be said to be 'designed' in the conventional sense of the term (administrative regulations, technical standards, machinery and installations) – 'emergent' structures. Aside from their characterization as 'complex',[13] the attribute 'emergent' is perhaps the most common one in LTS studies. This characterization does not in itself point to any more specific conditions for the growth of technically networked structures – except, of course, for the obvious periodization establishing LTSs as an emergent phenomenon of industrialized, Western societies. Again, emergence talk achieves, I think, welcome effects of LTS naturalization and unaccountability, presumably especially for later stages of seemingly out-of-control system dinosaurs.

Stories not told

Semioticians tell us that it is possible to say things only by not saying others. When Renate Mayntz noted that '(w)e may have been inclined to focus too much on types of hierarchical or centralized control, and have tended to frame the issue in terms of centralization/decentralization' (1991: 182), it suddenly became clear that there is a ubiquitous systems concept in the LTS discourse which accentuates hierarchy, although systems are hardly ever explicitly linked with this characteristic anywhere in these studies. But can the unstated always be derived so clearly from what is actually said?

I see four stories which have remained largely untold in LTS studies: the everyday users/uses of LTSs, the cultural and natural histories of LTSs, and the place of space (as opposed to phase) in LTS-research.

Users and the system

Imagine a couple of hundred pre-modern citizens, such as for example the types that appeared in French movie houses a few years ago in the film *Les Visiteurs*. And imagine that they are herded into a large AF Airbus. The usual take-off routines are played out and the plane starts. I cannot describe to you the terror and the fears and the reactions of the passengers. But I suggest that this aeroplane cannot be flown safely because the passengers have not acquired the many cognitive, emotional and practical disciplines that make us modern air travellers safe baggage. But we are hardly mentioned in LTS studies.

The main actors in LTS dramas – wherever they are written as stories with main actors – are either system builders in Hughes's sense, monopolistic utilities or government bureaucracies. But if research often forces the conclusion that technical systems do not evolve according to the intentions of their authors, could one not suspect that it is the recipients whose strategies determine what shape LTSs take? If LTSs do not follow any particular design, may not one assume that it is their users who influence their form of development? Interestingly, LTS users play no considerable part in the research examined here. Their possible influence is hardly investigated and is hidden conceptually behind metaphors like market or environmental control.[14]

Users are generally not treated as immediate objects of study. The implied users, the 'model users' of LTSs, are passive users. Their discipline, their active contribution, and by the same token their potential for resistance, are rarely made explicit. The co-production of system services is not explicated in LTS research. Users remain hidden, most of the time, behind abstract concepts such as demand or risk or regulation.

Two exceptions are MacKenzie (1991), in his study of supercomputing, and Rochlin (1991), in his reconstruction of the battle of the *Vincenne* in the Persian Gulf. And even MacKenzie and Rochlin operate here with a metaphor coming from social studies of small-scale technical systems: that of the 'hard-wired user'. The hard-wired user is the user in the machine, as it were, not the user before and after the machine. The notion is advanced that the machineries and technical systems under study are impregnated with a previously designed, built-in user model, forcing actual users to interpret their possibilities for action in a certain way. (But who are the end-users of a battleship?[15]) There is a mild technological determinism in the image of the 'hard-wired user' which can only be corroborated, or refuted, if and when the reciprocal impacts of systems on various user groups and users on various systems (with various wired-in users) are actually studied.

There is an issue in LTS debates where users feature more prominently. It is the issue of whether or not user control and user participation in late-modern communication networks such as the Internet differ from early modern technical systems such as energy generation systems and other classic infrastructures. One can often hear a thesis that while old systems such as energy or transportation systems are centrally controlled, more recent systems, such as, prototypically, the Internet, are decentralized and basically built up, managed and conducted by large numbers of loosely coordinated users. This underestimates, however, or misconstrues the part users play in classic systems and it overestimates and misconstrues the part they play in telecommunication systems such as the Internet. An energy generation and distribution system brings power to millions of everyday users who utilize this power for countless purposes: listening to Mozart, cooking spaghetti, reading the newspaper, heating their apartment and so on and so forth. In order to produce these goods they operate their household machines.

One could say that the centrally controlled system is programmed by users to produce their purposes. The same with road systems which are variously programmed by users with the help of their cars to produce their purposes: going to work, visiting friends. What happens in the Internet is not that different, apart from its technical scale: communication in the Internet relies on relatively few monopolistic corporate telephone networks, which are centrally controlled. Internet users and surfers program these networks with the help of their peripheral equipment for their own purposes. This is not to say that there are no differences between systems. In a user perspective, the main difference between old and new systems lies in the form of cooperation required from users. In the case of Internet, the user competences and resources required may be more visible and at least for a while more spectacular than in the case of today's electricity uses. But this does not give users more control.

Cultures and the system

Guided by the systems metaphor, LTSs are preferably construed as being systemic, rational and functional. Only recently have considerations of 'system culture' been gaining in importance, as for example in Rochlin's (1994) cross-national comparison of nuclear systems. But, to recall the example of Hughes's electrical systems: things may not have been quite so rationally planned, not to mention implemented and managed, back then around the turn of the century.

With his book *Electrifying America*, for instance, David Nye (1990) wrote something like a cultural history of the 'networks of power'. He argues that, quite at variance with the metaphors propagated by the inventors and operators of electricity at the beginning of the century, early applications of electricity were anything but embryonic systems, oriented towards well-centralized and well-integrated network structures and controlled by a disciplined elite of engineers and managers. Rather, Nye argues that incipient electrical technologies were experienced as 'sublime objects'. Electricity was considered as a thing of terrifying magnificence, whose significance, as presented to the awed and amazed masses by scientific magicians, lay in its capacity to frighten, but also to deepen and strengthen the thinking consciousness of the observer. According to Nye, the early electrical inventions in America were highly theatrical, spectacular, extravagant and dysfunctional.

Our present day LTSs 'evolved' and 'emerged', then, from a huge muddle of technically incompatible and competing enterprises, speculations and spectacles. Cultural acceptance was achieved through public displays, especially light shows, the 'conspicuous consumption' of cities. Of course there were also representations of electricity as functional and technically neutral, but they remained juxtaposed with representations of electricity and its possibilities as breathtaking, exciting, fantastic and magic. In short: electricity was subjected to an intensive cultural tug-of-war before it became effectively naturalized and its juice could begin to flow through all conceivable expressions of life in Western societies. Only in later times was this circumstance forgotten, partly because governments and corporations engaged in retrospective falsification in writing up their own official histories. It was a long way, then, from the scientific-technological miracle of the electric current to its effectively blackboxed technical applications in daily life, to a service reduced to a mere function and commercial product.

The emblematically sublime object of our time is the computer. Sherry Turkle, author of the much-quoted study on computers, *The Second Self*, describes these machines as the 'evocative objects' of the epoch: 'Under pressure from the computer, the question of mind in relation to machine is becoming a central preoccupation. It is becoming what sex was to the Victorian – threat and obsession, taboo and fascination' (Turkle 1984: 313). Today's counterpart of the nervousness of the electrical medium is clearly the autistic quality of communication in the frigid zones of electronic cyberspace.[16] And many researchers studying computer-based technical networks – that 'postmodern' type of LTSs – are too readily seduced by the sounds of the harp chords (or rather the synthesizer sounds) of a computer pop culture, tuned to simulacra, virtual reality and cyborgization. Retrospective LTS research may have had difficulty in adequately accounting for cultural processes; future-oriented research may take them into account too carelessly.

Natures, spaces, places and the system

Just as their cultural history is passed over, another blind spot in LTS studies tends to efface the natural histories of these systems. Because attention has almost exclusively been concentrated on the technical/social interface, the social/natural relationship in LTS development was largely ignored. Whatever happened to ecological issues in LTS discourse? Except for a remark about the bad weather conditions prevalent at the time of the Shuttle/Challenger disaster (Pinch 1991) and about the weather problem in flight control (La Porte 1988), the sparse allusions to natural conditions in the two weighty LTS volumes reviewed here take merely the negative form: 'Neither the British nor the Norwegian model could handle the challenges of the North Sea' (Andersen 1991: 46). Only recently did Thomas Hughes latch on to ecological LTS discourse in the fields of human geography and landscaping, bringing into the international research network the works of authors such as Cronon (1991) or Spirn (1984).

The ecological blind spot in LTS research is perhaps also related to the fact that the basic narrative mode is a linear, story-telling, not a lateral, space-mapping one. Even Giddens, the social theoretician who probably contributed most towards giving sociology a spatial dimension, meets this demand almost only for the spatial dimension of the nation state. Processes of spatial expansion, territorial appropriation and assimilation extending beyond regional and national boundaries are as characteristic of the growth of large technical systems, especially in connection with processes of urbanization, as are the celebrated stages of successive differentiation and integration over time. They have not been very carefully delineated in LTS research up until this time.

Foucault had a few interesting things to say about this, albeit without any direct reference to the problems of ecology and – one must now add – ecocracy. The great obsession of the nineteenth century was, he says, 'history, with its themes of development and of suspension, of crisis and cycle, themes of the ever-accumulating past'. Our epoch, he claims, will perhaps be above all the epoch of space: 'We are in the epoch of simultaneity . . . of the near and far, of the side-by-side, of the dispersed. We are at a moment . . . when our experience of the world is less that of a long life developing through time than that of a network that connects points and intersects with its own skein' (Foucault 1986: 22). And elsewhere he insists: 'A whole history remains to be written of spaces – which would at the same time be the history of powers . . . – from the great strategies of geopolitics to the little tactics of the habitat, . . . from the classroom to the design of hospitals, passing via economic and political installations' (Foucault 1980: 149). This applies to the spaces of technical networks too, from the micro-spaces of consumer technologies to the macro-spaces of global communication systems.

A 'reterritorialization' of LTS research seems indeed underway, as the latest round of the international LTS conferences indicates (this volume): both in terms of case-studies and in terms of conceptualizations, territory, space and place are given more thought than before. This accords with a growing emphasis in social studies of large technical systems on aspects of local embedding and the socio-cultural contextualization of disembedded technical and conceptual devices. There remains the other aspect, however: the business of system building is precisely to develop solutions for multiple applications in many places and to resist, as it were, premature closure. This seems to me the core competence of what Thomas Hughes has famously called independent inventor–entrepreneurs, or system-builders. It involves both a high awareness of the importance of local adaptation of system solutions and the ability to offer generic solutions which will then be turned into countless local applications by others: skilled disembedding of local solutions is a distinct achievement wherever new, and particularly international spaces are to be opened. It seems to me that 'reterritorialization' is a necessary step in understanding this feat of viably disembedding system solutions. How are 'universal services', i.e. services irrespective of place and space, possible?

Where to sail?

As Renate Mayntz noted at midpoint in the conference series: 'We have tried to draw . . . a very sketchy map of a new continent which needs considerable refinement before we can set sail for new shores' (Mayntz 1991: 181). The two conferences since have added detail without, however, changing the course of LTS research. Concluding this journey over some of the verbal carpets spread out by LTS research, one may wonder about the refinements as well as the new shores.

Narrating, interpreting, modelling?

Can a step be taken beyond the basically narrative, i.e., interpretative accounts of (even comparative) case studies in the general direction of systematic hypothesis-testing and model-building and ultimately quantitative comparisons? The historian William McNeill once differentiated between two paths in the social sciences observing that 'some historians (resort) to narrative in every case . . . (describing) surprising results from specific actions, and leaders combining old and new in a surprising future. Others (fall) back on numbers'. And he contends that the real intellectual issue is 'how to understand the interaction of the episodic but critical act with the underlying ebb and flow of numbers' (McNeill 1987: 110–11).

The others here are clearly sociologists and the like. Clearly, the reading public in general prefers historifying approaches. Linnda Capporael, for instance, concentrates her praise on the narrative chapters of Mayntz and Hughes (1988) in her discussion of this study: 'For readers who can surmount (or just skip) a lifeless and jargon-laden opening chapter, it is . . . useful . . . ' (Capporael 1990: 210). In fact, by far most of the studies referred to here use a combination of narrative, historical reconstructionist and model-like, counterfactual approaches. But the old problem – 'how to put numbers into stories', in the sense of quantitative modelling – remains unsolved in LTS studies, too.

An entirely different problem remains unsolved too, however, that of 'how to control one's interpretations'. Around the mid-1980s, when I first began to concern myself with LTSs, I looked for examples of stories and interpretations of large (too large) technology in literary fiction. I soon came across Alfred Döblin's *Berge, Meere und Giganten* (Mountains, Seas and Giants) which I found equally fascinating and repulsive. Döblin's expressionist envisioning of heroic, archaic, endlessly disruptive dramas of ever crazier technical schemes played out between ever less-familiar social forms seemed entirely unsuited to provide insights into the apparently irreversible upwardbound transformations and rationalizations of present-day and foreseeable large technical systems. Today, I see this work as a literary interpretation of the *longue durée* of LTSs, which has suddenly gained in plausibility in the wake of the reversals and revolutions of the recent past, especially in the Eastern European LTSs.

When I now read a latter-day American Döblin such as Neil Stephenson and his equally crazy post-cyberpunk visions in *Snow Crash* or *Diamond Age*, I certainly tend to overestimate the plausibility of his vision as much as I underestimated the heuristic value of Döblin's vision. In other words, my own interpretative patterns have shifted quite a bit in a rather short period of time in a way which I find hard to account for in the study of an empirical LTS. How can we reflexively account for such shifts and, more difficult still, for those heuristics that don't shift? One certainly should worry not only about our incapacity to put numbers into our stories; it would also be desirable to find a disciplined form for dealing with hermeneutical issues since interpretations are unavoidable and essential for modelling as much as for storytelling.

Forecasts

The category of 'large technical system' held out the promise of rendering an interesting conceptual status to the extended technical foundations without which modern societies would not be able to develop and maintain the breadth of their functional differentiations or their interactions over ever-increasing distances of social space and time. Should one regard the variability of the sociological discussion about LTSs as an indication that it is difficult to think of the LTS as a particular type of social system? Should one see it as an indication that the concepts of social systems theory, as they have entered the hybrid discipline of LTS research, do not provide the means for sufficiently sharp distinctions and enough requisite variety to enable consideration of all systematic and historical aspects? Whatever answer a reader might give to these questions, LTS research should and will contribute to our better understanding of one of the central issues in an anthropology of industrial and superindustrial societies: how to make and maintain control and order at very great distances, particularly at a stage or in an age when familiar centralized, hierarchical and national-territorial forms of governance have seemingly become overextended and can no longer manage the trans-border and even global processes LTSs have occasioned and made possible.

After this wonderfully long sentence, my final, more specific predictions are: ambitious, sufficiently broad and necessarily comparative empirical investigations under way in a number of places will make their way from research fields into books and journals. The interlacing of different LTSs will attract more attention. Consideration of other approaches to the study of technology – which place less faith in the systems metaphor – will add interesting overtones to LTS research. Analyses of LTSs from the user perspective will follow. Cultural studies of LTSs will appear on the bookshelves. Place and space will gain interest over temporal considerations. Historians and sociologists of technology will put aside their quibbles. Now and then, the study of LTSs will even profit from the injection of small doses of irony and self-reflection.

The high variability of LTS discourse will, if anything, increase. This should be seen as a resource, however, rather than something to be remedied. It would be too easy to reduce the apparent paradoxicality of LTS discourses – emergence or design? equilibrium or crisis? system or rhizome? – to a series of two-by-twos into which concrete instances can then be sorted. The paradoxicality resides irreducibly in the so-called systems, not in our feeble attempts to understand them. The best way of dealing with it is to understand how system actors – builders, users – deal with it.

Notes

* This chapter is partially based on 'Reden über grosse Technik', published in Hans-Ulrich Derlien, Uta Gerhardt, Fritz Scharpf (eds) (1994) *Systemrationalität und Partialinteresse, Festschrift für Renate Mayntz*, Baden-Baden: Nomos, 453–90.

1 Although I will take the stance of the outsider/analyst in the following pages, I consider myself an insider/participant of this community, having participated in all the conferences except Sydney and having organized a separate series of seminars with a different circle of German scholars, which has resulted in two more books on LTS (Braun and Joerges 1994b, Joerges 1996c).

2 I am painfully aware that this procedure in no way does justice to the authors of the original texts, not only because my interpretations and attributions necessarily diverge from the authorial intentions, but because this form of quoting violates subtle linguistic contexts. Largely excluded from these intertexts are meta-texts aiming at higher level abstractions and generalizations on the basis of empirical LTS research done elsewhere (such as Mayntz 1988, 1993, Joerges 1992, Braun 1994, also Gras 1994, Grundmann 1994, Joerges and Braun 1994, Summerton 1994a).

3 In the subsequent intertexts I apply, for purposes of readability, two simplifying conventions: (i) Instead of referring to concrete systems or system types, I use the generic word 'system' throughout; (ii) differences in tense forms are eliminated and all quotes are put in the present tense. Quotations are separated by en rules.

4 The one move to discuss categorical issues in depth comes from Beckman (1994), albeit not in the context of an empirical LTS study.

5 The same holds for the concept of complexity (Joerges 1996b).

6 Even though Charles Perrow attended the first international meeting.

7 Interestingly, this can even be achieved for artificial systems; on the naturalization of (computer) technology, see Joerges (1991).

8 For further discussions see von Meier 1994: 211–26, Radkau 1994: 71–83, Kubicek 1994.

9 With a few exceptions (Hughes 1989, Rochlin 1991, Bucholz 1994), military technology, state-technology par excellence, plays no role.
10 Regrettably I do not have, at the point of writing this, a text from which I could quote.
11 It would be intriguing to consider the stages/phases debate and its variations (four stages, five stages and their dramaturgical transformations) in the light of a structuralist narratology such as was first influentially developed by Vladimir Propp 1968.
12 For further references to the design-issue see Salsbury 1994: 159, Usselman 1994: 101–6, Schneider 1994: 78–81, Rochlin 1994: 231–55, Kornwachs 1994: 430–4, von Meier 1994: 219–24, Abbate 1994: 200–2.
13 For a more detailed account of the 'discourse of complexity' in LTS research see Joerges (1996b).
14 For more on the role of large-scale technical background and 'warranty' systems in everyday applications of small-scale consumer technologies, see Braun (1989).
15 It seems a bit implausible to apply the interpretation of the hard-wired user to the victims of military attacks.
16 For a classic literary text on this see Gibson (1986).

References

For purposes of clarity, the very frequent references to four edited volumes have been abridged in this bibliography. The volumes are: Braun and Joerges 1994b, La Porte 1991, Mayntz and Hughes 1988, Summerton 1994b.

Abbate, Janet (1994) 'The Internet challenge: conflict and compromise in computer networking', in Summerton 1994b: 193–210.

Andersen, Svein S. (1991) 'The external control of large technical systems: offshore oil production safety and health regulation in Great Britain and Norway', in La Porte 1991: 43–60.

Beckman, Svante (1994) 'On systemic technology', in Summerton 1994b: 313–32.

Bertho-Lavenir, Catherine (1988) 'The telephone in France 1879 to 1979: national characteristics and international influences', in Mayntz and Hughes 1988: 156–77.

Braun, Ingo (1989) 'Technische Infrastrukturen der Konsumarbeit', *Zeitschrift für Umweltpolitik und Umweltrecht* 4: 353–75.

—— (1994) 'Geflügelte Saurier. Zur intersystemischen Vernetzung grosser technischer Systeme', in Braun and Joerges 1994b: 447–99.

Braun, Ingo and Joerges, Bernward (1994a) 'How to recombine large technical systems: the case of European organ transplantation', in Summerton 1994b: 25–52.

—— (eds) (1994b) *Technik ohne Grenzen*, Frankfurt am Main: Suhrkamp.

Bucholz, Arden (1994) 'Armies, railroads, and information', in Summerton 1994b: 53–70.

Capporael, Linnda R. (1990) 'Review of Mayntz and Hughes, *The Development of Large Technical Systems*' *Contemporary Sociology* 19, 2: 210.

Caron, François (1988) 'The evolution of the technical system of railroads in France from 1832 to 1937', in Mayntz and Hughes 1988: 69–103.

Cronon, William (1991) *Nature's Metropolis: Chicago and the Great West*, New York and London: W.W. Norton.

Ekardt, H.-P. (1994) 'Unter-Gestell. Die bautechnischen Fundamente grosser technischer Systeme', in Braun and Joerges 1994b: 166–211.

Foucault, Michel (1980) 'Questions on geography', in C. Gordon (ed.) *Power/Knowledge: Selected Interviews and Other Writings 1972–1977*, New York: Pantheon.

—— (1986) 'Of other spaces', *Diacritics* 16: 22–7.

Galambos, Louis (1988) 'Looking at the boundaries of technological determinism: A brief history of the U.S. telephone system', in Mayntz and Hughes 1988: 135–53.

—— (1991) 'A view from economic history', in La Porte 1991: 177–80.

Gibson, William (1986) 'Burning chrome', in William Gibson, *Burning Chrome*, New York: Ace Books.

Giddens, Anthony (1989) *The Consequences of Modernity*, Oxford: Polity Press.

Gras, Alain (with Sophie L. Poirot-Delpech) (1994) *Grandeur et dépendance: Sociologie des macro-systèmes techniques*, Paris: Presses Universitaires de France.

Grundmann, Reiner (1994) 'Über Schienen, Strassen, Sand und Perlen. Grosse technische Systeme in der Theorie sozialer Systeme', in Braun and Joerges 1994b: 501–44.

Heinze, G. Wolfgang and Kill, Heinrich H. (1988) 'The development of the German railroad system', in Mayntz and Hughes 1988: 105–34.

Hughes, Thomas P. (1987) 'The evolution of large technological systems', in Wiebe Bijker, Thomas Hughes and Trevor Pinch (eds) *The Social Construction of Technological Systems: New Directions in the History and Sociology of Technology*, Cambridge (MA): MIT Press.

—— (1989) *American Genesis: A Century of Invention and Technological Enthusiasm*, New York: Viking.

—— (1991) 'A historical overview', in La Porte 1991: 185–8.

Joerges, Bernward (1991) 'Images of technology in sociology: computer as butterfly and bat', *Technology and Culture* 31, 2: 203–27.

—— (1992) 'Große technische Systeme: Zum Problem der Maßstäblichkeit und Vergrößerung in des sozialwissenschaftlichen Technikforschung', in G. Bechmann and W. Rammert (eds) *Jahrbuch Technik und Gesellschaft*, Frankfurt am Main: Campus, 41–72.

—— (1996a) '"Große technische Systeme" oder: was ist groß und was ist klein?', in B. Joerges (ed.) *Technik – Körper der Gesellschaft*, Frankfurt am Main: Suhrkamp, 145–71.

—— (1996b) 'Large technical systems and the discourse of complexity', in Lars Ingelstam (ed.) *Complex Technical Systems*, Stockholm: Affairsliteratur AB.

—— (ed.) (1996c) *Körper–Technik: Aufsätze zur Organtransplantation*, Berlin: edition sigma.

Joerges, Bernward and Braun, Ingo (1994) 'Grosse technische Systeme: erzählt, gedeutet, modelliert', in Braun and Joerges 1994b: 7–49.

Kornwachs, Klaus (1994) 'Steuerung und Wachstum. Ein systemtheoretischer Blick auf grosse technische Systeme', in I. Braun and B. Joerges (eds) *Technik ohne Grenzen*, Frankfurt am Main: Suhrkamp, 410–45.

Kubicek, Herbert (1994) 'Steuerung in die Nichtsteuerbarkeit. Paradoxien in der

Entwicklung der Telekommunikation in Deutschland', in Braun and Joerges 1994b: 107–65.

La Porte, Todd (1988) 'The United States air traffic system: increasing reliability in the midst of rapid growth', in Mayntz and Hughes 1988: 215–44.

—— (ed.) (1991) *Social Responses Large Technical Systems*, Dordrecht: Kluwer.

Lévy-Leboyer, Maurice (1988) 'The French electrical power system: an inter-country comparison', in Mayntz and Hughes 1988: 245–62.

MacKenzie, Donald (1991) 'Notes toward a sociology of supercomputing', in La Porte 1991: 159–76.

Mayntz, Renate (1988) 'Zur Entwicklung technischer Infrastruktursysteme', in R. Mayntz, B. Rosewitz, U. Schimank and R. Stichweh (eds) *Differenzierung und Verselbstaendigung: Zur Entwicklung gesellschaftlicher Teilsysteme*, Frankfurt am Main and New York: Campus.

—— (1991) 'A view from the social sciences', in La Porte 1991: 181–4.

—— (1993) 'Grosse Technische Systeme und ihre gesellschaftstheoretische Bedeutung', Koelner *Zeitschrift für Soziologie und Sozialpsychologie* 45, 1: 97–108.

Mayntz, Renate and Hughes, Thomas P. (eds) (1988) *The Development of Large Technical Systems*, Frankfurt am Main and Boulder (CO): Campus and Westview Press.

Mayntz, Renate and Schneider, Volker (1988) 'The dynamics of system development in a comparative perspective: interactive videotext in Germany, France and Britain', in Mayntz and Hughes 1988: 262–98.

McCloskey, Donald N. (1985) *The Rhetoric of Economics*, Madison (WI): University of Wisconsin Press.

McNeill, William (1987) 'Reflections by William McNeill', in U. Svedin and B. Aniansson (eds) *Surprising Futures*, Stockholm: Swedish Council for Planning and Coordination of Research.

Meier, Alexandra von (1994) 'Integrating supple technologies into utility power systems: possibilities for reconfiguration', in Summerton 1994b: 211–30.

Mulkay, Michael and Gilbert, Nigel (1992) 'What is the ultimate question? Some remarks in defense of the analysis of scientific discourse', *Social Studies of Science* 12: 309–19.

Nye, David E. (1990) *Electrifying America: Social Meanings of a New Technology*, Cambridge (MA): MIT Press.

Oster, Clinton V. Jr (1991) 'The aviation safety commission: a task force approach to the evaluation of large technical systems', in La Porte 1991: 127–42.

Pinch, Trevor (1991) 'How do we treat technical uncertainty in system failure? The case of the space shuttle Challenger', in La Porte 1991: 143–58.

Propp, Vladimir (1968) *The Morphology of the Folktale*, Austin (TX) and London: Texas University Press.

Radkau, Joachim (1994) 'Zum ewigen Wachstum verdammt? Jugend und Alter grosser technischer Systeme', in Braun and Joerges 1994b: 50–106.

Rochlin, Gene J. (1991) 'Iran Air flight 655 and the USS Vincennes: complex, large-scale military systems and the failure of control', in La Porte 1991: 99–126.

—— (1994) 'Broken plowshare: system failure and the nuclear power industry', in Summerton 1994b: 231–61.

Salsbury, Stephen (1988) 'The emergence of an early large-scale technical system: the American railroad network', in Mayntz and Hughes 1988: 37–68.

Salsbury, Stephen (1991) 'Facing the collapse of the Washington power supply system', in La Porte 1991: 61–98.
—— (1994) 'The Australian electric power industry and the politics of radical reconfiguration', in Summerton 1994b: 141–62.
Schneider, Volker (1991) 'The governance of large technical systems: the case of telecommunications', in La Porte 1991: 19–42.
—— (1994) 'Multinationals in transition: global technical integration and the role of corporate telecommunication networks', in Summerton 1994b: 71–92.
Sen, Amartya (1992) 'On the Darwinian view of progress', *London Review of Books* 14, 21 (5 November): 15–19.
Spirn, Ann Winston (1984) *The Granite Garden: Urban Nature and Human Design*, New York: Basic Books.
Staudenmaier, John M. (1985) *Technology's Storytellers: Reweaving the Human Fabric*, Cambridge (MA): MIT Press.
Summerton, Jane (1994a) 'Introductory essay: the systems approach to technical change', in Summerton 1994b: 1–24.
—— (ed.) (1994b) *Changing Large Technical Systems*, Boulder (CO): Westview Press.
Thomas, Frank (1988) 'The politics of growth: the German telephone system', in Mayntz and Hughes 1988: 179–213.
Turkle, Sherry (1984) *The Second Self. Computers and the Human Spirit*, New York: Simon and Schuster.
Usselman, Steven W. (1994) 'Changing embedded systems: the economics and politics of innovation in American railroad signaling, 1876–1914', in Summerton 1994b: 93–118.
Wagner, Gerald (1994) 'Vertrauen in Technik', *Zeitschrift für Soziologie* 23, 1: 145–57.
Weingart, Peter (1989) '"Grosstechnische Systeme": ein Paradigma der Verknüpfung von Technikentwicklung und sozialem Wandel?', in Peter Weingart (ed.) *Technik als sozialer Prozess*, Frankfurt am Main: Suhrkamp.
—— (1991) 'Large technical systems, real-life experiments, and the legitimation trap of technology assessment: the contribution of science and technology to constituting risk perception', in La Porte 1991: 5–18.

GENERAL INDEX

INDEX OF AUTHORS CITED